The View from the Oak

The View from the Oak

The Private Worlds of Other Creatures

by Judith and Herbert Kohl   Illustrated by Roger Bayless

Sierra Club Books / Charles Scribner's Sons

San Francisco / New York

The View from the Oak was edited and prepared for publication at The Yolla Bolly
Press, Covelo, California, under the supervision of James and Carolyn Robertson
during the fall and winter of 1976-1977. Production staff: Sharon Miley, Loren
Fisher, Marie Dern, Gene Floyd, Jay Stewart, Evelyn Swift.

The Sierra Club, founded in 1892 by John Muir, has devoted itself to the study and
protection of the nation's scenic and ecological resources—mountains, woodlands,
wild shores and rivers. All Club publications are part of the nonprofit effort the
Club carries on as a public trust. There are some 50 chapters coast to coast, in
Canada, Hawaii and Alaska. Participation is invited in the Club's program to enjoy
and preserve wilderness everywhere. Address: 530 Bush Street, San Francisco,
California 94108.

MANUFACTURED IN THE UNITED STATES OF AMERICA

LIBRARY OF CONGRESS CATALOGUING IN PUBLICATION DATA

Kohl, Judith.
The view from the oak.

Bibliography : p. 110
SUMMARY: Attempts to enable us to view the world of ticks, flies, birds, jelly fish,
and other animals through their senses, not our own.
1. Zoology—Miscellanea—Juvenile literature.
2. Animals, habits and behavior of—Miscellanea—Juvenile literature.
3. Senses and sensation—Miscellanea—Juvenile literature.
4. Observation (Psychology)—Miscellanea—Juvenile literature.
[1. Animals—Miscellanea. 2. Senses and sensation—Miscellanea. 3. Observation
(Psychology)—Miscellanea.]
I. Kohl, Herbert, joint author.
II. Title

ISBN 0-684-15016-6

ISBN 0-684-15017-4 pbk.

1 3 5 7 9 11 13 15 17 19 MD/C 20 18 16 14 12 10 8 6 4 2   3 5 7 9 11 13 15 17 19 MD/P 20 18 16 14 12 10 8 6 4

QL49.K66   591'.028   76-57680

This book was inspired by Jacob von Uexkull's essay *A Stroll Through the World of Animals and Men: A Picture Book of Invisible Worlds.* Though our work is quite different from von Uexkull's essay, his writings have given focus to our efforts and his ideas have helped form our thinking.

Acknowledgments

Grateful acknowledgment is made to the authors and publishers of the following for granting us permission to reprint material originally published elsewhere.

From Chapter One, "The Bee-Hunters of Hulshort," in *Curious Naturalists* by Niko Tinbergen, © 1958 by Niko Tinbergen, Basic Books, Inc. Publishers, N. Y. Reprinted by permission of Doubleday & Co., Inc.

Instinctive Behavior, Part 1, "A Stroll through the World of Animals and Men" by Jacob von Uexkull. N. Y.: International Universities Press, Inc. Copyright © 1957 by International Universities Press, Inc. Reprinted by permission of International Universities Press, Inc.

Wild Heritage by Sally Carrighar. Boston: Houghton Mifflin Company. Copyright © 1965 by Sally Carrighar. Reprinted by permission of Houghton Mifflin Company.

Motivation of Human and Animal Behavior by Konrad Lorenz and Paul Leyhausen. Copyright © 1973 by Litton Education Publishing, Inc. Reprinted by permission of Van Nostrand Reinhold Company.

The Penal Colony by Franz Kafka. Copyright © 1948 by Schocken Books Inc. Renewed © 1975 by Schocken Books Inc. Reprinted by permission of Schocken Books Inc.

The Magic of the Senses by Vitus B. Dröscher. London: W. A. Allen Company. Copyright © 1969 by E. P. Dutton. Reprinted by permission of E. P. Dutton.

King Solomon's Ring by Konrad Lorenz. N. Y.: Thomas Y. Crowell Company. Copyright © 1952 by Thomas Y. Crowell Company. Reprinted by permission of Thomas Y. Crowell Company.

Contents

PART I

Worlds

Worlds

We now know that there is not one space and one time only, but that there are as many spaces and times as there are subjects, as each subject is contained by its own environment which possesses its own space and time.
Jacob von Uexkull
Theoretical Biology

Our dog Sandy is a golden retriever. He sits in front of our house all day waiting for someone to come by and throw him a stick. Chasing sticks or tennis balls and bringing them back is the major activity in his life. If you pick up a stick or ball to throw, he acts quite strangely. He looks at the way your body is facing and as soon as you throw something, he runs in the direction you seemed to throw it. He doesn't look at what you threw. His head is down and he charges, all ears. If your

A dog's best friend is his nose.

stick lands in a tree or on a roof, he acts puzzled and confused. He runs to the sound of the falling stick and sometimes gets so carried away that he will crash into a person or tree in the way as he dashes to the place he hears the stick fall. As he gets close, his nose takes over and smells the odor of your hand on the stick.

Once we performed an experiment to see how sensitive Sandy's nose really was. We were on a beach that was full of driftwood. There was one particular pile that must have had hundreds of sticks. We picked up one stick, walked away from the pile and then threw it back into the pile. It was impossible for us to tell with any certainty which stick we had originally chosen. So many of them looked alike to us that the best we could do was pick out seven sticks which resembled the one that had been thrown.

We tried the same thing with Sandy, only before throwing the stick we carved an X on it. Then we threw it, not once but a dozen times into the pile. Each time he brought back that stick. Once we pretended to throw the stick and he charged the driftwood pile without noticing that one of us still had the stick. He circled the pile over and over, dug out sticks, became agitated but wouldn't bring another stick. It wasn't the shape or the size or look of the stick that he used to pick it out from all the others. It was the smell we left on the stick.

It is hard to imagine, but for dogs every living creature has its own distinctive smell. Each person can be identified by the smell left on things. Each of us gives off a particular combination of chemicals. We can detect the smell of sweat, but even when we are not sweating, we are giving off smells that senses finer than ours can detect.

The noses of people have about five million cells that sense smell. Dogs' noses have anywhere from 125 to 300 million cells. Moreover, these cells are closer to the surface than are cells in our noses, and more active. It has been estimated that dogs such as Sandy have noses that are a million times more sensitive than ours. Clothes we haven't worn for weeks, places we've only touched lightly indicate our presence to dogs. Whenever Sandy is left alone in the house, on our return we find him surrounded by our sweaters, coats, handkerchiefs, shirts. He surrounds himself with our smell as if to convince himself that we still exist and will return.

His ears are also remarkable. He can hear sounds that humans can't and at distances which are astonishing. It is hard for us to know and understand that world. Most of us don't realize that no two people's hands smell the same. Our ears are not the tuned direction finders his are. It takes a major leap of the imagination to understand and feel the world the way he does, to construct a complicated way of dealing with reality using such finely tuned smell and hearing. Yet his world is no more or less real than ours. His world and ours fit together in some ways and overlap in places. We have the advantage of being able to imagine what his experience is like, though he probably doesn't think too much about how we see the world. From observing and trying to experience things through his ears and nose we can learn about hidden worlds around us and understand behavior that otherwise might seem strange or silly.

The environment is the world that all living things share. It is what is—air, fire, wind, water, life, sometimes culture. The environment consists of all the things that act and are acted upon. Living creatures are born into the environment and are part of it too. Yet there is no creature who perceives all of what is and what happens. Sandy perceives things we can't, and we perceive and understand many things beyond his world. For a dog like Sandy a book isn't much different than a stick, whereas for us one stick is pretty much like every other stick.

There is no one world experienced by all living creatures. Though we all live in the same environment, we make many worlds.

Jacob von Uexkull, who wrote *A Stroll Through the World of Animals and Men* 45 years ago, was one of the first people to try to imagine the world as lived and perceived by insects and animals. He created the word *Umwelt* to describe the world around a living thing as that creature experiences it. It's easy to understand why he had to create a new word instead of using one that already existed. There is no word in English or German that stands for the world of an animal as that animal experiences it. Words such as "world," "experience," "nature," or "reality" won't do. *Umwelt* expresses something quite different—it stands for organized experience that is not shared by all creatures, but is special to each creature. The word *Umwelt* is German and the plural should be *Umwelten*. However umwelt is used in this book as a new English word and therefore the English plural umwelts will be used, also it will not be capitalized or italicized.

Think, for example, of an ant and a bee in a field of blooming flowers. The ant lives in a colony buried under the ground in a small part of the field. Through its entire life it probably never travels from one end of the field to the other. In its world flowers in bloom, budding plants, trees, bushes are all obstacles to climb or bypass. Their differences are of no importance in the ant's life and aren't perceived. The ant spends most of its day scurrying about searching for food which it brings back to the colony. It is extremely sensitive to slight vibrations in the earth and works as part of ant teams which it communicates with by touching antennae or stomping on the earth and creating vibrations. Imagine being an ant and working your way through particles of dirt, small rocks, grass, and flowers as humans would go through a forest or field strewn with boulders. The world of an ant is rich with detail and

Ant (*Solenopsis xyloni*)

variation. However, many things going on in the field do not enter its world.

Bees and ants do not usually cross paths. The field of blooming flowers is perceived in a special way by bees. They can smell the perfumes given off by flowers over great distances and can distinguish flowers by their scents. Some flowers are rich in pollen and others have very little. Bees choose the rich flowers first and as soon as they pick up their scent, fly towards them. As they get close to the flowers, they use visual as well as scent clues to find the flowers and estimate their numbers. Bees' eyes are particularly suited to identify and distinguish different flowers. According to experiments performed by Karl von Frisch, bees cannot distinguish the shapes shown here.

These shapes, which resemble buds rather than blooming flowers, all appear as dark roundish shapes in the bees' world. On the other hand, bees have no difficulty telling the difference between these two flower shapes.

In the world of the bee the field appears full of indefinite circles or of various flower shapes. The world is either blooming flower or other. These pictures illustrate the field as a complex environment shared with the ant and other creatures, and as it appears in the world of a bee.

In von Uexkull's terms the ant and bee described here share the same environment but live in different umwelts.

Learning to Observe

It is difficult to understand the umwelt of creatures that have different senses and sizes than ours. It is tempting to interpret animal life in terms of human images. An ant may be small in our minds but a large or normal-sized being in its world. A bird that seems to be fluttering about in pain and confusion to us may merely be flirting with another bird. A menacing buzz of a bee or hornet may be nothing more than the sound they usually make when moving about.

We have to be cautious and work our way slowly into the worlds of animals. Patience is necessary, and silence. You have to learn how to be in the presence of animals without disturbing them, and to be able to distinguish slight differences between animals of the same species.

Suppose you try to observe the life of an ant. One way to go about it would be to capture an ant and watch how it functions in a jar. This would be a bad choice because an ant cannot live and perform its ordinary life functions without being part of a community of ants. It would die of aloneness.

Another way to study the ant would be simply to look at it. That is not as easy as it sounds. To follow a single ant with your eyes and not lose it under a leaf or get it confused with the other ants around it takes intense concentration. You can start by observing for a few minutes at a time and slowly build up your stamina. You have to heighten the awareness of your senses at the same time as relaxing. Observing animals is almost a form of meditation.

To get an idea of how involved a person can get in the lives of animals, listen to Niko Tinbergen, one of the greatest students of animal behavior, describe how he uncovered the life of a bee-killing wasp:

On a sunny day in the summer of 1929 I was walking rather aimlessly over the sands, brooding and a little worried. . . . While walking about, my eye was caught by a bright orange-yellow wasp. . . . It was busying itself in a strange way on the bare sand. With brisk, jerky movements it was walking slowly backwards, kicking the sand behind it as it proceeded. The sand flew away with every jerk. . . . I stopped to watch it and soon saw that it was shovelling sand out of a burrow. After 10 minutes of this,

it turned around, and now facing away from the entrance began to rake loose sand over it. In a minute the entrance was completely covered. Then the wasp flew up, circled a few times around the spot describing wider and wider loops in the air, and finally flew off. . . . I expected it to return with prey within a reasonable time and decided to wait. . . . Sitting down on the sand, I looked round and saw that I had blundered into what seemed to be a veritable wasp town. . . .

I had not to wait long. I saw a wasp coming home . . . it was carrying a load, a dark object about its own size . . . a Honey Bee. . . . I watched these wasps at work all through that afternoon and soon became absorbed in finding out exactly what was happening in this busy insect town. . . . This day, as it turned out, was a milestone in my life. For several years to come I was to spend my summers with these wasps, first alone, then with an ever-growing group of co-workers, most of them zoology students of the University of Leiden. . . .

Settling down to work, I started spending the wasp's working days (which lasted from 8 A.M. till 6 P.M. . . .) on the "Philanthus plains" as we called this part of the sands as soon as we found out that Philanthus triangulum Fabr. was the official name of this bee-killing digger wasp. . . . An old chair, field glasses, notebooks, and food and water for the day were my equipment.

What to Look For

When observing animals we must try to give ourselves over to their experience and imagine worlds as foreign as any that can be found in novels or science fiction. It helps to focus on three aspects of experience that all animals have to deal with and organize to make their lives coherent. One aspect is locating oneself in space. A second is growth and change, that is, time; and a third is developing a way to respond to threat and friendship, or to food, i.e., to tone and mood. Space, time, and response are central aspects of the lives of all animals. Through an understanding of how these aspects are organized by particular animals, it is possible to get an idea of how they feel and function. This means that we have to disorient ourselves—to give up our own usual sense of how time flows and what distance means as well as our sense of what feels threatening or friendly. To become close to other worlds means giving up our own for a while.

The easiest way into the spatial, temporal, and responsive worlds of animals is through an understanding of the way their senses work. What vibrations does an ant respond to? Does light play any role in the world of a flatworm? What is an eel's response to sound? Such questions can help us to understand the umwelts of different creatures.

PART II

Having a Place
in the World

Having a Place
in the World

A central aspect of the lives of all animals is how they experience space—that is, how they experience where they are and where other things are located. Without a sense of space animals would be unable to hunt, to have a sense of how to return to their homes or how to escape when they're pursued. There would be no way to find a mate, to build a nest, to deposit eggs, or have young in protected places. Space is one of the major coordinates in the life of all creatures, though we must be careful to understand that all animals don't experience space in the same way.

Imagine how the sense of space might develop in people. Most of us have a good idea of where our bodies end and the rest of the world begins. But think of a newborn baby.

The first weeks of a baby's life are full of learning about where food comes from, what it tastes like, feels like. The baby learns to associate a particular sound with a particular person or animal or machine and that the sound comes from "out there," although it may not know how far away it actually is. A baby quickly learns that a sound it can make inside itself, crying, often brings food and comfort.

At first the baby has no sense of where the food or the comforting person comes from. The space world of the baby is probably no larger than its crib. Later, when babies begin to crawl, they develop an idea of how far away different places in their world are and where people and pets in their homes spend time. But imagine where the baby thinks a stranger who walks into the house after ringing the door bell comes from. What does the other side of the door represent as space for the baby?

Imagine the sense of space a baby would have if he or she couldn't feel or could feel so finely that the movement of electrons was perceptible. Imagine what space would be like for a baby who could feel and smell but not see or hear.

A normal baby's sense of space beyond the door is not far from our sense of space "out there" beyond the doors of our experience. In the 15th century, people thought the world was flat. When Columbus set out on his voyage to the New World, people thought he was crazy because he was certain to fall off the edge of the earth. The world of these people was as flat as a pancake. But the world for the early explorers wasn't flat. They believed that if they sailed far enough in one direction and kept sailing, they'd end up where they started—because the earth is round. The same people who thought that the earth is flat also believed that the sky is like a big tent, that somewhere up there is a kind of lid over the earth that holds the stars and the moon and the sun. Now that we have seen pictures from space, we know that the moon isn't suspended from a heavenly "ceiling." But we still have trouble imagining what outer space is really like. Our umwelt is changing as we learn more.

Close your eyes and pass one of your hands from left to right in front of them. Think about where right and left begin. Where is the middle of your body as you feel it with your eyes closed? Try more than once. Do you feel anything when your hand passes your nose?

Now with the palm of one hand facing the ground close your eyes again and pass your hand up and down in front of your face. Somewhere between your eyebrows and your upper lip you probably felt the place where the sensation of up changes to down and down to up.

Now as a third experiment, raise your palm to ear level, face it forward, close your eyes, and move your palm forward and back. Where does front begin? Back? Most people sense a change near the ear, others as far forward as the tip of the nose. When some people try this experiment, they experience a slight dizzy sensation whenever their hands cross these imaginary lines. But are they imaginary?

If you were to draw a picture of the three planes where you experience the difference between up and down, right and left, and front and back, these planes would intersect roughly at a spot in your head where the organs of balance, the semicircular canals, are located. These organs help us know when we are falling or tilting or are well balanced. They function in our bodies like a gyroscope does on an ocean liner— the gyroscope provides information about when the ship is listing too dangerously to one side by keeping itself spinning upright no matter what movement of the ship is caused by the ocean. It is an internal navigator that helps the ship keep oriented in space.

A gyroscope basically consists of a spinning wheel mounted in a movable frame. You can get an inexpensive gyroscope in most toy stores and use it to perform amazing tricks and experiments with

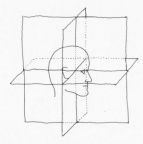

Where up and down, right and left, and front and back meet

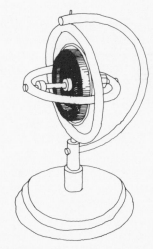

Gyroscope

balance. While spinning, it can balance on your fingertip even if you move your hand about. It can also balance on the edge of a drinking glass, on a spoon or long stick. If you try gently to push it down, it won't fall, but will continue spinning and balancing at what seems an impossible angle. The axle of the gyroscope always points to the same direction no matter how the object it is spinning on is moved. That's how it is used in navigation on a ship. No matter where the ship moves, the axle of its gyroscope points in the same direction. It is like an internal North Star which can be used as a fixed point to locate oneself in space. You can observe how this works with an inexpensive gyroscope and a toy boat. Put the boat in a full bathtub and put a spinning gyroscope on the boat. Then push the boat in a curved path and observe what happens to the gyroscope.

The semicircular canals in our ears seem to function as our gyroscopes. They provide us with a sense of up and down, right and left, and front and back, however we may be thrown around. Of course they don't always work—too much spinning or alcohol or anesthesia throws them off, and there are times that we literally don't know where we are.

The semicircular canals consist of three canals filled with tiny hairs and liquid. When the liquid is sloshed about, the hairs that are usually dry get wet and signal the brain that we are off balance. If the rest of our body works reasonably well, our muscles then act to correct the imbalance. If the canals are shaken violently, they make us feel dizzy.

Usually we don't think much about this. We correct small changes in our balance almost automatically. We hold our bodies differently when walking up steep hills than when climbing down the stairs. It is probably only cases of extreme imbalance and violent movement that make us nauseous or dizzy and we become aware of the need to get back

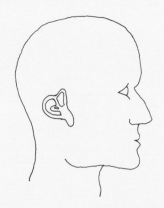

The gyroscope in your head—the semicircular canals

into balance. For the rest, our inner gyroscope functions smoothly and automatically and enables us to orient ourselves without too much effort.

There are other animals whose balance mechanisms are similar to ours. Fish, for example, have three semicircular canals and orient themselves in water much the way we do in air.

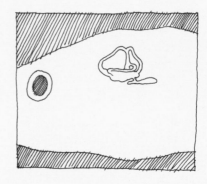

Other insects have leg hairs that seem to function similarly to the hairs in our semicircular canals. Ants have tiny hairs at all their joints, not only on their legs but at the joints of their bodies and heads. When the ants' bodies are straight and their legs are down in a resting position, these little hairs are not being rubbed or pressed on by any of the joints. But as soon as the ant lifts a leg, the leg hairs are touched, and the ant knows something about the relationship between that leg and the ground. Should he turn his abdomen sideways, he knows he is headed in a new direction. When he lowers an antenna to feel a pebble in his path, the hairs at the antenna joint tell him the height of the pebble in relation to himself. The ant's sense of space is three dimensional, but it is touch, not sight, that is so important in its umwelt.

To get a feel of what this might be like, set up a small obstacle course with cartons, large rocks, a few chairs, old tires, etc. Then blindfold yourself and try to move through the course crawling on your hands and knees. One problem will probably develop quickly. You'll have to find a way to protect your head from crashing into the obstacles. Without the use of your eyes, there is need for some other sense organ to move ahead of the body and provide information about possible dangers.

If you're not afraid of looking silly to your neighbors and friends, you could get even closer to the ant's world by tying two thin flexible plastic rods to your head, thus making your own antennae. As you move your head about, you'll begin to interpret the vibrations that come from the antennae and find them as useful as canes are to blind people.

Trying to imagine the worlds of animals or actually becoming part of their worlds, often leads people to behave in ways that seem bizarre to uninformed neighbors or casual spectators. Konrad Lorenz, who won a Nobel Prize in biology for his work with animals, described a moment when he must have appeared to be insane to a group of strangers. He had raised a group of mallard ducks to believe that he was their mother. As he describes it, in order to get the ducklings to follow him he had to move along squatting low, making ducklike sounds.

This was not very comfortable; still less comfortable was the fact that the mallard mother quacks unintermittently. If I ceased for even the space of half a minute from my melodious "Quahg, gegegegeg, Quahg, gegegegeg," the necks of the ducklings became longer and longer corresponding exactly to "long faces" in human children—and did I then not immediately recommence quacking, the shrill weeping began anew. As soon as I was silent, they seemed to think that I had died, or perhaps that I loved them no more: cause enough for crying! The ducklings, in contrast to the greylag goslings, were most demanding and tiring charges, for, imagine a two-hour walk with such children, all the time squatting low and quacking without interruption! In the interests of science I submitted myself literally for hours on end to this ordeal.

So it came about, on a certain Whit-Sunday, that, in company with my ducklings, I was wandering about, squatting and quacking, in a May-green meadow at the upper part of our garden. I was congratulating myself on the obedience and exactitude with which my ducklings came waddling after me, when I suddenly looked up and saw the garden fence framed by a row of dead-white faces: a group of tourists was standing at the fence and staring horrified in my direction. Forgivable! For all they could see was a big man with a beard dragging himself, crouching, round the meadow, in figures of eight, glancing constantly over his shoulder and quacking—but the ducklings, the all-revealing and all-explaining ducklings were hidden in the tall spring grass from the view of the astonished crowd.

The World as a Yo-Yo

Up-down, right-left, and front-back are not central orientation points in the lives of all creatures. For a clam, right-left and front-back are not as important as up-down. The clam has to know the difference between up and down as it digs its way in and out of its sandy or muddy world. It is like a yo-yo, moving up and down with the tides. The group of animals to which clams belong, mollusks, has balancing organs which are of the same type as ours, although theirs are much simpler and give the clam much less information. Within the animal is a cavity that is lined with tiny hairs, and resting in the cavity is a small round stone that gravity keeps in a down position. Depending on which way the clam tilts or is washed by the waves, certain of the hairs are touched by the stone as it rolls in the cavity, and messages are picked up by the clam's nervous system. If the clam is off balance, it then has information to move quickly so that the stone is resting once again on the hairs that indicate balance. Thus the clam can tell whether it is parallel to the ground or at a slant. When it is parallel, then digging will take it up or down according to its needs. To imagine the umwelt of a clam, it is necessary to emphasize this up-down orientation and forget about left and right and front and back.

You could make a model of the balancing organ of a clam using a narrow plastic box, a heavy metal bead, some string and wire, bells, and batteries. The model would look like the drawing on this page.

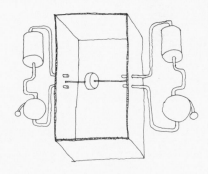

Model of a clam's balance mechanism

If the box is upright, indicating that the clam is moving either up or down but not on an angle, nothing will happen. If the box is tipped, the bead goes over to the side causing a bell to ring. This is a signal to correct the position of the box. If it is overcorrected, the bead will slip over to the other side. When both bells are off, that indicates that the box is in an upright position.

A Flat World

Imagine living inside a piece of paper and having to orient yourself and move on that surface. There would be left-right and front-back orientation, but up and down would not exist. Such a world would be two dimensional. Edwin Abbott in his book *Flatland* tries to create such a world. In the first chapter the narrator who is a citizen of a two-dimensional world tries to explain his world:

I call our world Flatland, not because we call it so, but to make its nature clearer to you, my happy readers, who are privileged to live in Space.

Imagine a vast sheet of paper on which straight Lines, Triangles, Squares, Pentagons, Hexagons, and other figures, instead of remaining fixed in their places, move freely about, on or in the surface, but without the power of rising above or sinking below it, very much like shadows— only hard and with luminous edges—and you will then have a pretty correct notion of my country and countrymen. Alas, a few years ago, I should have said "my universe": but now my mind has been opened to higher views of things.

In such a country, you will perceive at once that it is impossible that there should be anything of what you call a "solid" kind; but I dare say you will suppose that we could at least distinguish by sight the Triangles, Squares, and other figures, moving about as I have described them. On the contrary, we could see nothing of the kind, not at least so as to distinguish one figure from another. Nothing was visible, nor could be visible, to us, except Straight Lines. . . .

Place a penny on the middle of one of your tables in Space; and leaning over it, look down upon it. It will appear a circle.

But now, drawing back to the edge of the table, gradually lower your eye (thus bringing yourself more and more into the condition of the inhabitants of Flatland), and you will find the penny becoming more and more oval to your view; and at last when you have placed your eye exactly on the edge of the table (so that you are, as it were, actually a Flatlander) the penny will then have ceased to appear oval at all, and will have become, so far as you can see, a straight line.

The same thing would happen if you were to treat in the same way a Triangle, or Square, or any other figure. . . .

There are some animals whose spatial worlds approximate Flatland. Think, for example, of a water strider, an insect that glides on the surface of the water. It lives on top of the thin film that creates tension on the surface of water. Water striders can't swim, and if they happen to break this tension and become submerged, they drown.

Water striders have thousands of tiny hairs on their legs (like the hairs in our semicircular canals) which help them stay on top of the water and turn and stop as they skim the surface. The active world of this animal is almost entirely two-dimensional; the third dimension, however, is a source of danger. Fish come from below and birds from above to catch the water strider. But its activity is two-dimensional— on the surface of the water. It is not even clear whether the water strider is aware of its predators or if its darting movement on the surface provides all the protection water striders need as a group in order to survive.

It's possible to experiment with water striders without hurting them or disrupting their environments in a damaging way. Near our house is a small pond with a few goldfish and innumerable water insects. Water striders can be distinguished from the other pond insects by their long graceful legs and by the way they seem to move along on a cushion of air. We caught a few and put them in a wide-mouthed jar full of pond water. Then we experimented with approaching the striders by lowering different objects above them. They didn't respond to our approach. But if we touched the water, however gently, they immediately began to glide around the jar in an agitated manner. This seemed to indicate that their world was indeed two dimensional.

You can easily catch water striders and experiment with them yourself. Keep them in a jar full of pond water and be sure the jar is kept in a cool place. If you plan to keep them for any length of time, they will have to be fed live insects such as flies or mealworms. It's best to let them go after a while. There are many questions you can try to answer: How do water striders respond to darkness? What do they do when you tap the bottle with a spoon? Do they respond if you approach them from below? Can you touch the water so softly or quickly that they don't respond?

Space without Dimension

Think of a perfectly round ball. It has no front or back, no right or left. Now imagine the ball sailing through endless space, occasionally bumping into things and bouncing off them but not going anywhere in particular. There is no up or down in such space. The ball, if it could have any experience, would live in a space without dimension or direction. There are very small animals that live in a space similar to that.

One-celled organisms such as parameciums (which also have tactile hairs) orient themselves in a way that could be described as retreat-attack. They have no front or back, no up or down side, no right or left. They live in water and move with the currents, spinning about, bumping into things from all angles. Some have small hairs that function as paddles. There is nothing, however, in their bodies that could be called faces or fronts. They don't move forward or backward—they simply move. Nevertheless, they do respond to things in their world. Once they make contact with something, they either absorb it if it is edible or attempt to retreat if it is not. Their space is structured by the simplicity of their experience. When your movement is fairly random and your responses limited, space doesn't assume distinct dimensions. There is no such thing as right side up or wrong side up in such a world. Think, for example, of how an amoeba might orient itself—not only does it not need three-dimensional balance; it has no stable shape.

Parameciums can be found in most pond water and observed by using a 50- or 100-power microscope. If you watch them for 10 or 15 minutes, you will see how they spin and collide and retreat seemingly at random. By putting a fine pin in the water, you can experiment with how they respond to pressure. The lower the magnification you can use (50x instead of 100x), the better sense of space in the parameciums' umwelt you will get, since the smaller magnification gives you a larger though less detailed area to observe.

Underwater creatures

Can You Hear Space?

Have you ever played pin the tail on the donkey or blindman's buff? These games begin by spinning you around, shaking up your semicircular canals, and trying to disorient you. Your eyes are closed, and you have to locate a person or target without the use of sight. Pin the tail on the donkey depends upon how well you remember how many times you've been spun around and on how well you can resist getting dizzy and confused. Blindman's buff has another element. You can hear people's voices and their movements and can make a reasonable guess of their location from these sounds. Though we don't often think about it, we locate things in space through hearing as well as sight and touch. Our umwelt is built up by all of our senses and sometimes it takes eliminating one or another of them to tell how much the remaining senses contribute.

Our sense of sound in space is poorly developed compared to that of many animals. Part of the fun of blindman's buff is that it is more difficult for us to locate things in space using our ears than using our eyes. For example, if something makes a sound behind us, we don't always know how far away it is, nor can we always tell in which direction it lies. Although we can know whether it is to the left or right, our ears can barely distinguish between sounds directly in front or in back of us. Close your eyes and ask a friend or two to move as quietly as possible either directly in front or behind you and about ten feet away. They shouldn't tell you whether they are in front or behind you. Instead, when they reach the positions they choose, have them clap loudly once or ring a bell or bang two pieces of wood together. Does the sound come from in front of you or from behind you?

Your eyes have limits just as your ears do. Hold your head still and look as far to the left and then as far to the right as you can. That distance defines the limits of your field of vision at a given moment. We do not have 360-degree sight. Yet in the umwelt of many animals 360-degree sight and hearing exist. They see and hear with a precision we find only in instruments people have developed to imitate animal

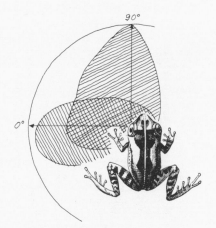

A frog's field of vision

senses. In fact, there is a branch of engineering devoted to designing machines that function like the senses of animals. It is called bionics. The development of radar was based on the hearing of bats and the development of cameras and bombsights on the eye of an eagle. Yet even the best of machines can't approximate the sensitivity and range of animal senses, and even the most aware scientists have only been able to guess at the umwelts of the animals they study.

In the case of hearing we have to imagine refining and intensifying our human capacities and eliminating the central role played by sight in the human umwelt. For example, imagine an owl on a moonless night. It can catch its prey, not so much because of its large eyes, but because an owl's ears can determine the spot where a mouse is as accurately as we can know where the lamp in front of us is.

The ability of whales and dolphins to navigate and communicate by sound is phenomenal. Whales can send messages to each other over hundreds of miles and can find each other on the open sea by swimming towards each other's sounds. While swimming at top speed, dolphins estimate the speed and location of moving objects according to the sounds they make and the way these sounds are reflected.

Marco Polo is a water game that can give you an idea of what the world of a dolphin must be like. It is best played by four or more people in a small swimming pool. One person is chosen as "it" and has to stand on the side of the pool with closed eyes and count to ten. The other people jump in the pool and spread out. After getting to ten, "it" jumps in the water and shouts out "Marco." Throughout the game the player who is "it" has to keep his or her eyes closed. As soon as "Marco" is shouted out, all the other players have to answer by shouting "Polo." The object of the game is for "it" to catch one of the other players who then takes a turn being "it."

Dolphins navigate and communicate by sound.

During the game the players can keep moving, "it" shouting "Marco," everyone answering "Polo" faster and faster. That is sound navigating—moving yourself on the basis of sound made by something that is also moving. It's amazing how fast you can react in Marco Polo until you almost forget that you're not using your eyes at all.

Learning about the World by Touching It

Try this experiment with a friend. Put a number of smooth-surfaced objects in a carton. Choose, say, a piece of paper, a ball, a cereal box, a marble, a metal bottle top, a glass, a milk carton, and a piece of planed wood. Then close your eyes and have your friend choose any object out of the box. Put your middle finger on the object and don't move your finger at all. Can you tell what object it is? Try again and again. Most likely every object will feel the same unless one, like the metal, feels colder than the other. After doing this try again, only this time move your finger slowly over the surface. At what point do you feel that you know which object you're touching?

The sense of space built up by touch is not static. It has to do with movement over and around surfaces. It is an intimate form of contact with the environment, and yet one that depends upon how finely tuned your senses are. What is a smooth surface for us can be a rugged and irregular world for a tiny insect.

There are some animals that are very dependent upon touch for getting around in space. Rats and cats, for example, rely heavily on their whiskers and other protruding hairs while hunting and moving around. They function without trouble in the dark and are quite capable of getting around even if they are blinded. Moles are nearly blind yet feel their way through an elaborate series of tunnels and passages.

Bees seem to have a feeling for some shapes and dimensions. Before laying eggs, the queen bee measures the cells in the hive with her tactile abdominal hairs. If the cell is small, she fertilizes an egg and drops it in. This egg will become a worker bee who will tend to the hive, gather honey, and raise young bees. If the cell is larger, the queen doesn't fertilize the egg but merely drops it in. That egg will become a drone, a fertile male whose primary function is to mate with specially raised fertile female bees in order to create new queen bees.

A dramatic example of how an animal can navigate using touch is provided by the gerbil which is active at night and can jump enormous distances at amazing speed. If the ground is bumpy or the land irregular, the gerbil could easily crash in the dark. However, during its jump the gerbil points two hairs nearly as long as its body straight downwards. These hairs keep touching the ground during the jumps and warn it of holes, rocks, and other obstacles. It can then change its course in midair if it feels danger.

In order to understand how animals like the gerbil or bee or mole manage, we have to study what they use that constitutes such a major component of life as movement in space. It is through the study of the senses and the cues animals pick out of the environment as well as how they move and act that we can begin to imagine and understand umwelts that are very different from ours. As you approach an animal to understand what its experience is like, ask yourself what senses it depends upon; then imagine what it would feel like to perceive and organize the world with that information. Also, perform experiments in your imagination. Look at a tennis ball, for example, and imagine getting small enough to crawl on its surface. Then imagine yourself drilling through the ball and falling into its center. Pull away and become bigger and bigger until the ball is no longer visible. Then try being another object, an apple you are eating through, a seedpod carried by the wind, a cat hunting in a field. These exercises in imagining new worlds are not strictly speaking scientific, but they are ways of preparing yourself to encounter unfamiliar worlds and accept unexpected results. They help provide the openness that must inform any serious and respectful study of animal life. Often in trying to figure out the umwelt of an animal the most unexpected worlds open up, and we must be prepared to learn from them.

Senses beyond Imagination

How do rattlesnakes find out where their prey is? This question puzzled biologists for quite a while. If the snake was blinded, it had no trouble spotting and catching its prey. The same was true if its nostrils were blocked. Neither sight nor smell nor touch nor hearing played a role in detecting food. All the known senses were eliminated, and for a while the conclusion had to be that the rattlesnake had mystical powers beyond our comprehension. That interpretation, though not helpful in specific ways, wasn't far from the truth. The rattlesnake does have a third eye—two of them in fact.

There are two pits above a rattlesnake's mouth that are sensitive to very small changes in temperature. We sense temperature too, but whereas we have about three heat-sensitive points per square centimeter on our skin, the rattlesnake has more than 150,000 on its pit organ which is located just under its eyes. With that organ it can distinguish differences of a few tenths of a degree centigrade and can pick out things the way a bat can with sound. Even lizards disguised as leaves, motionless, with little scent or sound aren't safe. The snake senses the change in temperature caused by their body.

A rattlesnake "sees" by sensing changes in temperature.

39

Imagine an umwelt where space was defined by temperature differences. Many distinctions we perceive would not be different in that umwelt. On the other hand, things we perceive to be the same would be significantly different. We don't see or usually even sense temperature differences in a heated room. Yet to a rattlesnake the room would be full of variations; but objects with the same temperature would not appear as distinct. Try to draw a temperature map of a room full of people attending not to shape, form, smell but only differences in heat. The blurring of outlines normally known and the creation of distinctions that we usually don't notice give a good idea of how we have to stretch beyond the familiar to understand the experience of animals.

Using Your Eyes

There are a number of animals, people included, whose main orientation in space is built up through vision. However, there are different organs of sight and many ways of organizing the information these organs send to the brain. There is no single visual space—it depends upon how you see, what size you are, and what you're looking at. A passage in Jonathan Swift's *Gulliver's Travels* shows this better than we can. Gulliver is having dinner with a giant Brobdingnagian farmer:

> *In the midst of dinner, my mistress's favourite cat leapt into her lap. I heard a noise behind me like that of a dozen stocking-weavers at work; and turning my head, I found it proceeded from the purring of this animal, who seemed to be three times larger than an ox. . . . The fierceness of this*

creature's countenance altogether discomposed me. . . . But it happened there was no danger; for the cat took not the least notice of me. . . .

When dinner was almost done, the nurse came in with a child of a year old in her arms, who immediately spied me, and began a squall . . . to get me for a plaything. . . . The nurse to quiet her babe made use of a rattle, which was a kind of hollow vessel filled with great stones, and fastened by a cable to the child's waist: but all in vain, so that she was forced to apply the last remedy by giving it suck. I must confess no object ever disgusted me so much as the sight of her monstrous breast, which I cannot tell what to compare with, so as to give the curious reader an idea of its bulk, shape and colour. It stood prominent six foot, and could not be less than sixteen in circumference. The nipple was about half the bigness of my head, and the hue both of that and the dug so varified with spots, pimples and freckles, that nothing could appear more nauseous. . . . This made me reflect upon the fair skins of our English ladies who appear so beautiful to us, only because they are of our own size, and their defects not to be seen but through a magnifying glass, where we find by experiment that the smoothest and whitest skins look rough and coarse, and ill coloured.

I remember when I was at Lilliput, the complexion of those diminutive people appeared to me the fairest in the world; and talking upon this subject with a person of learning there, who was an intimate friend of mine, he said that my face appeared much fairer and smoother when he looked on me from the ground, than it did upon a nearer view when I took him up in my hand and brought him close, which he confessed was at first a very shocking sight. He said he could discover great holes in my skin; that the stumps of my beard were ten times stronger than the bristles of a boar, and my complexion made up of several colours altogether disagreeable: although I must beg leave to say for myself, that I am as fair as most of my sex and country, and very little sunburnt by all my travels. On the other side, discoursing of the ladies in that Emperor's court, he used to tell me, one had freckles, another too wide a mouth, a third too large a nose; nothing of which I was able to distinguish.

We rely on our eyes to give us a huge amount of information about the space around us. Our eyes are not very accurate mechanisms. Much of the work is done by the brain interpreting what information it gets from our eyes. A group of people were once subjected to an experiment where they had to wear glasses that distorted everything they saw— made things appear upside down, wavy. Surprisingly enough, after a time everyone adjusted quite well to all this and actually began to "see" things as they always had. When they eventually took off the glasses, the world around them appeared as distorted as it did when they first put on the glasses! And, they had to go through another period of re-adjustment before everything appeared right side up again.

When light passes through the lens of our eyes, it falls on the retina inside the eyeball which acts somewhat like the film inside a camera. The retina is made up of many separate cells; each registers part of the "picture" that we are looking at and, through connecting nerves, these separate pictures are sent to the brain and put together into a total picture. Our eyes move constantly when we look at anything. You can only see a spot the size of your thumbnail held at arm's length very clearly at any moment, because only one small part of the retina has enough cells to give us a really clear picture. The remaining retinal cells are not dense enough to give us anything but a blurry image.

Our eyes almost automatically move all over the field of vision and pick up hundreds of separate clear pictures which are constantly being put together by the brain. What we get is a constantly changing picture of the parts of the whole, not unlike the impressions that make up our

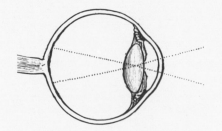

Your eye works like a camera.

identification of someone in the blindman's buff game. If you stare at something too long, its image will actually disappear because each sensitive spot on the retina wears out quickly. It must not have light focusing on it for very long if it is to register a picture. Fortunately, the retina is not like camera film in that it renews itself very quickly.

The number of cells on the retina which receive images from the outside world varies among animals. By examining the density of visual cells in the eyeballs of various animals it has been determined, for example, that a lion probably sees about as clearly as we do, and a fly's eye is capable only of what to us would be blurred vision, something like what we see at the edges of our visual field. A hawk or osprey, on the other hand, sees his world so acutely that we would have to look through binoculars that magnify our visual space by eight times to equal what the hawk sees.

Imagine Sandy, our golden retriever, standing on the beach. He would look different in the umwelts of different animals. On the next page is how he might appear to a fly, a person, and an osprey.

There are other eyes that work on principles very different from the ones described so far. We have to keep our eyes moving and scanning all the time to get coherent images. The frog, for example, has eyes that don't move. They are like a blank screen, and there is no image registered by them unless something is moving past. Frogs register moving objects only when they are moving towards them. They don't see anything if an insect or bird is moving away. If you sneak up on a frog in a pond at night and suddenly turn on a flashlight, the frog will remain still as if it's frozen in its tracks. What's actually happening is that it hasn't seen anything. Put your finger in front of the flashlight and move it towards the frog and the frog will be gone as fast as it can. The frog's eye works just like the photoelectric cell that is used to open doors in

supermarkets or count people or cars passing a certain point. In fact, the cell was designed on the model of a frog's eye.

On the Horizon

Just as there are limits to how clearly any species of animal can see, there are limits to how far it can see. This farthest plane that an animal can see can be called its horizon.

The location of the horizon is not rigidly fixed. It depends upon where we're standing, how we feel, and most of all, on the limits of our senses. However, the farthest we can extend our horizon provides a limit to the world we live in. For people the visual horizon is only a temporary limit—for us the whole earth is open to be travelled and experienced. We know we can change our position and create new horizons. But for an ant or moth or dog? What is their horizon and how do we come to know it? There are some hints we can get by watching the way animals behave and studying their senses. Flies, for example, don't simply circle around a light, but interrupt their flight suddenly whenever they've flown about half a meter or so away from the light. At that point they zero back in as if the light is about to become invisible to them and they have to stay within that radius.

The farthest plane or horizon seems to frame the experience of all seeing animals. Jacob von Uexkull described the situation beautifully:

We may . . . picture all the animals around us, be they beetles, butterflies, flies, mosquitos or dragonflies that people a meadow, enclosed within soap bubbles, which confine their visual space and contain all that is visible to them. . . . The fluttering birds, the squirrels leaping from branch to branch, or the cows that browse in the meadows—all remain permanently surrounded by their soap bubbles which define their own space.

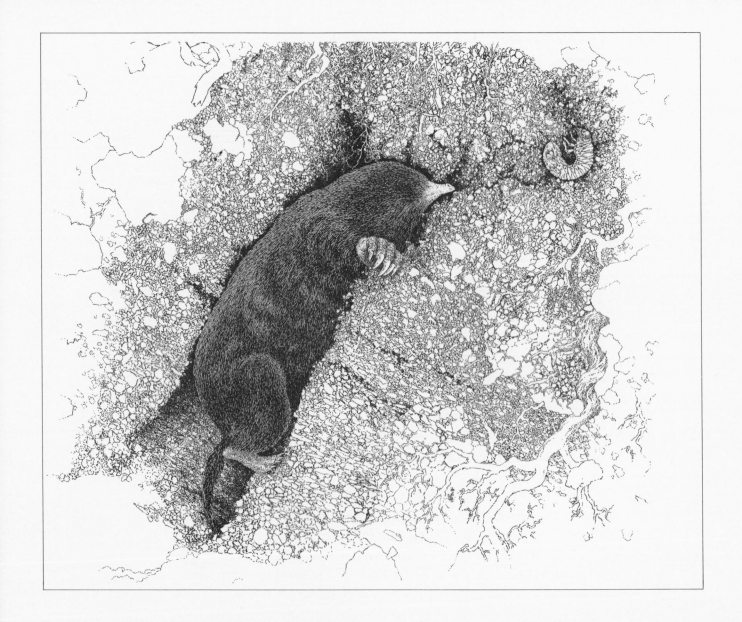

Only when this fact is clearly grasped shall we recognize the soap bubble which encloses each of us as well. . . . There is no space independent of subjects.

The mole lives in a series of tunnels under the ground. During the course of a day's hunting, the mole travels through this maze many times, pouncing on whatever smaller animals might get caught in or near the tunnels. Though it is nearly blind, it knows all the twists and turns in the tunnel and can smell prey that is in the earth up to a distance of five or six centimeters from the tunnel. In the midst of this network of tunnels the mole builds itself a cave home padded with dry leaves. This is where the mole drags its prey, has its young, and sleeps.

The whole complex of tunnels is its territory and it will fight any other mole that comes near. If the mole is used to finding prey in a certain tunnel and the tunnel is destroyed, the mole can entirely rebuild it.

All dirt might smell and feel and look alike to us. But for the mole each part of its territory has a particular density, smells of all the creatures that do or have lived there, and has its own texture. To understand how a mole knows its territory, the next time you go to the beach, pick up a pinch of sand and take it home. Then look at the individual grains of sand under a 20- or 40-power magnifier. Are there any two grains alike? What is the range of colors? of shapes? of densities? Imagine being just a bit larger than the grains of sand. A pinch of sand would be as varied and complex as the dirt around the mole's tunnel is for the mole.

The weak-eyed mole not only can tell different parts of its tunnel; if a whole series of tunnels collapses, the mole can rebuild new tunnels

Home and Territory

that closely resemble the old ones. All of this means that using smell and touch and sound the mole can orient itself in space, find its way home, and orient other points in space well enough to find them with no visual clues.

Many animals build or find homes for themselves and define a territory which they defend from any other animal of the same species who tries to invade their territory. However, not every animal has both a home and a territory. For spiders, their webs are both their homes and their territories. For the bee, the hive is its home, but the field where it gathers pollen is open to bees from many different territories. It has a home and a range where it seeks food but no territory. A housefly buzzing around for food or a place to drop its eggs seems to have neither a home nor a territory.

Home and territory are made by animals, not found in the environment. They are what von Uexkull calls umwelt problems, products of the actions of the subjects dependent on the way they hunt and on the way their senses organize the environment. North American brown bears mark their territory by finding tall pine trees at the limits of the turf, then they stand as tall as they can and rub off the bark of the tree with their backs and snouts. This is a signal to other bears to keep away. It plants sensory cues in the environment that are probably meaningful only to another bear. Such cues would not exist in the world of a snake or ant or hawk.

If any piece of woodland or forest were considered, there could be many different territorial maps drawn from it. Even a small patch of land, no larger than a mole's domain, contains the territories of many different animals. Each creature has its own way of dividing the space it lives in. Each map would be like a political map indicating lines of defense and attack drawn up by the inhabitants of the land.

In the illustration on the opposite page are the homes of four creatures: a mole, a fox, a thrush, and an owl. Can you find those homes? Imagine each creature setting out from home to explore its territory. Do they ever encounter each other?

48

Sometimes territories are quite complex. For example, for hawks there is a neutral space between their nest and their hunting ground. The birds will not attack prey within this safe area, and many smaller songbirds take advantage of this and make their nests in the neutral zone. People think the reason for this zone is so that birds of prey don't mistakenly attack their own young when they just leave the nest and are learning to fly.

The same meadow that contains birds of prey might also contain badgers, raccoons, robins, foxes, ground squirrels, owls, moles, lizards, toads, and turtles—and they all have territories. Yet the territory of one kind of animal usually does not exist in the world of other species of animals.

However, animals take their territories as seriously as we take the boundaries of our cities and states and countries. The male stickleback fish, which is about three inches long, builds a nest. After the nest is finished, strange transformations take place. The male begins to glow. He changes color from a dull blackish green to red, blue green, and emerald. In this condition the male both searches for a female and defends the territory around its nest from other male sticklebacks. If another male approaches, he'll be chased away, possibly rammed. As soon as the chase reaches the limits of the stickleback's territory, the defender becomes less fierce. He isn't interested in a fight to the death over territory. In fact, at the boundary line both male sticklebacks abandon the chase altogether and stand on their heads, nibbling on the sand. It is almost as if they acknowledge jointly the boundary line and make peace.

Fierce behavior within a territory, and especially near a nest or prime hunting place, and a weakening of anger on the boundaries are not just characteristic of fish. Sally Carrighar, in her book *Wild Heri-*

tage, which is filled with interesting portraits from animal life, describes how families of gibbons go about defending their territories in groups:

It was the family's habit to go about through their territory of thirty-five acres or so each day, stopping in whatever trees the fruits, leaves, and young growing tips were at their most tempting stage. Meanwhile, they would make sure that no neighboring gibbons were trespassing.

Hearing the chattering up ahead, the father . . . took off abruptly. What he suspected was true: a family of neighbors were climbing into his fig tree, whose fruits were just then becoming deliciously ripe. Swinging in through the air, the father shrilled out a protest. The trespassing gibbons paused. Standing upright on a thick low branch of the fig tree, the father warned them away in a voice like an attack. The other family yelled back and, since there were six of them, their noise could overwhelm his. The sons joined in the father's bellowing, and the other gibbons reluctantly went down the tree. The father and sons followed them to the ground and on to the boundary of the properties. Once more in their own territory, the six stood and hurled taunts. As he answered, the father's voice was becoming more mild, not forgiving yet but no longer outraged. He had won, and before many minutes he might exchange friendly calls with these neighbors, for a gibbon's anger is quick to subside.

Dogs do the same thing. Sandy is surrounded by other male dogs on our street. There are a German shepherd, a Lab, several poodles, and a number of mongrels. All the male dogs have marked out their territory. If a new male dog comes into the neighborhood, there are a number of battles over turf. The first time we saw Sandy go after a Lab and the two of them growl, bare fangs, and go after each other's neck, we were terrified. But the fight, though serious, was not murderous. At one point the Lab turned on its back and exposed its neck to Sandy. He accepted the sign of surrender and walked away. Sandy almost always wins on his territory and loses on another dog's turf. The stickleback, Sandy, and most of us are alike in that we are fiercest in defending our own homes and fight with less conviction the farther away we get from home.

It is interesting to try to draw some territorial maps. Study a small tract of land. Where do the birds, squirrels, and spiders nest? How far do they move during the course of a day? How far apart are the nests of the same type of animals?

Even in the city it is possible to study animal territory by considering dogs and cats. Where is each dog most fierce? What are the boundaries where dogs leave each other alone? Where do they leave their marks? Does every dog have the same territory? And is there a special spot each dog has where it rests and if approached growls and shows its teeth?

Similar questions can be asked of cats, mice, wasps, and, of course, people. Studying home and territory is a good specific way to understand the importance of the umwelt, of what the subject puts into the environment.

It might be interesting to set up a system for observing a particular human territory and actually spending several days or even weeks charting and writing down in a notebook whatever happens within the space selected to observe. There are any number of places to choose from: playgrounds, schoolyards, bus stops, parking lots, street corners, public swimming pools, and beaches, and, of course, individual homes and apartments. Just be sure to choose a place that can be visited frequently and at many different times of day. In a house, for instance, which places are used by one person exclusively—a bed, bedroom, comfortable chair, the kitchen stove perhaps? Which are shared simultaneously—kitchen or dining room table, TV set? And which are used in sequence—the bathroom? Which places are considered off limits to which members of a family, which to outsiders, which are, like hallways, merely pathways to other more clearly defined areas? What happens when a person or pet violates another's declared territory?

However, it is not an easy task to study the territory of most familiar animals. Paul Leyhausen, whose ethological studies have led him to the observation of the cat family, once tried with a collaborator to collect a day-and-night record of the encounters between three free-ranging farm cats. He said:

It was an impossible task. To follow a single cat around day and night without losing sight of it, and keep a complete record of all its movements, encounters, etc., requires at least three well-trained, physically fit, and inexhaustible observers, plus a lot more equipment than we could command at the time. We carefully selected an isolated farmhouse situated in a clearing in a very hilly region. There were two resident cats, and another one in a farm some 600 yards away. Sufficient data were collected for only one of the residents, to form a picture from which we hope no essential feature is missing: but even this was not complete.

There are several different ways of studying animal life. One way is to isolate an animal in a laboratory and do things to it. Then one merely observes how the animal responds and reports on those responses. This approach which is usually called behaviorism concentrates on what is done to an animal and how the animal responds. What the animal thinks or how the animal organizes experience is not of immediate concern. Behaviorists usually take animals out of their natural environments and isolate them in laboratories. They are not concerned

The Importance of Understanding Space

with the rhythms of an animal's life, with its social world, or with its sense of space or time. The animal is put in an experimental situation. It might be taught to run through a maze, avoid an electric shock, or push a lever to get food. In any case, what is studied is the behavior of the animal in the experimental situation. It makes no difference whether one is studying a pigeon or dog or rat—what is important to behaviorists is the animal's reaction which is isolated from the rest of its life and its experience.

There is a completely different, more respectful, approach to the study of animal life which is called ethology. Ethology is concerned precisely with how an animal experiences the world and organizes the environment. It too is concerned with behavior but in the context of an animal's overall experience. Ethologists acknowledge that one can never get into the head of another living being, can never experience the world exactly as animals do. However, there are clues we can use to help build up an intelligent model of other forms of experience. These clues are provided by animals' senses and by the range of things in the environment that these senses make it possible for the animal to perceive. The clues are also provided by observations of how animals act in their natural environments and by occasional laboratory experiments that attempt to reproduce natural conditions in a controlled setting.

Ethologists do not try to isolate behavior from experience. On the contrary, they use behavior to develop an integrated view of different animals' lives. When they study the senses, for example, they are not concerned solely with how a sense works or what it registers. They try to go beyond the immediate data of the senses and understand how that data is organized so that an animal can integrate it and act upon it. For example, when an eagle sees a certain form in the grass, it not only recognizes the image of something grey but sees a mouse, begins to hover, and pounces exactly on that spot. It does not fling itself through space randomly. The information provided by the senses is organized in some overall ways that create a world of experience, an umwelt.

Some central aspects of that umwelt are the way an animal perceives its body, the way it locates itself with respect to other things in the environment, the way it moves, and the range of movement it has—that is the way space is organized. The space in an animal's umwelt is not an objective fact for everyone to observe. Often there are not visible territorial markers or actual horizons. The world is neither the way an eagle sees it or a person or a mouse. The space felt by a mole is no more or less real than that sensed by a rattlesnake's third eye. Space is created by each animal and at the same time frames each animal's experience. It is a characteristic of the umwelt, not a common element of the experience of all creatures.

But what of real space, physical space? Isn't there some absolute space in addition to all the different spatial worlds created by animals? The answer is no, not even in physics. According to the special theory of relativity, there is no such thing as absolute space. In order to talk about size and distance we have to create measures (like yards, meters, and rods) or to develop reference points (like the sides of our bodies or the direction of a given star). We can then talk about space in relation to these units. But these units themselves cannot be determined in any absolute way. They are starting points which enable us to use spatial concepts such as large-small, up-down, and near-far.

Henri Poincaré, a famous mathematician who lived in the 19th century, devised an exercise in imagination to help people understand the relativity of measures. Imagine that one night while you were asleep everything in the universe became a thousand times larger than before. Remember this would include electrons, planets, all living creatures, your own body, and all the rulers and other measuring devices in the world. When you awoke could you tell that anything had changed? Is there any experiment that you could make to prove that some change had occurred?

According to Poincaré there is no such experiment. In fact, it would be meaningless to say the universe grew larger since the word "larger" means larger in relation to something else. However, when talking about the whole universe, there is no something else to compare it to.

In physics, space is defined relative to the measures people create. Within the animal world space is relative too. It depends upon how animals sense things and organize this experience. Space in the umwelt is the psychological equivalent of time in the theory of relativity as we will see in the next part of this book.

PART III

Time and Change

Time and Change

Look at a clock. It is easy to follow the movement of the second hand as it sweeps around the face. Turn your attention to the minute hand. It is harder to see it move, though if you look patiently and intensely you can see it move too. The same is true of the hour hand, but that is even more difficult to see. Imagine a clock that had two more hands—one that would make a trip around the face once every 100 years and another that made the same circuit once every 1/100 of a second. The 100-year hand would move so slowly that it is unlikely that any of us could perceive it and pretty certain that very few of us would live long enough to experience a single rotation of that hand. The 1/100-second hand also causes problems for our eyes. It would speed around the clock so fast that the most we could perceive is a blur. There are some things that change too slowly for our eyes and others that change too quickly.

Think of the speed of a flower or tree growing, the wings of a hummingbird, a hawk circling in the sky, the electrons that make up your body, the feet of a horse running at top speed, the growth of a chicken embryo, light, a slug. We can perceive some of these things as they develop. Others are too fast or too slow, and we need to use our imaginations to understand them. A sense of the passage of time is built up out of the changes we perceive. If we had no senses, could experience no change, and never changed ourselves, the idea of time would not exist in our worlds. That is why many people who have tried to write about God or about possible causes of life and the universe conclude that these first causes have to be beyond and outside of time—changeless.

The Rhythm of Life

All living beings experience change and are subject to it as well. Over 2,500 years ago Galen, a Greek natural scientist, wrote about bean plants that folded their leaves at night and raised them during the day, day after day after day. Centuries later a number of scientists observed that when they compared beans grown in a garden with those grown in total darkness and at a constant temperature in the laboratory, the up and down movement of the leaves of both groups of plants was identical.

The botanist Carolus Linnaeus observed a large number of flowers that opened and closed at regular times of the day and was able to design a flower clock. Watching the opening and closing of the flowers on Linnaeus' clock, one can tell the time of day within a half hour.

We could probably make up a version of Linnaeus' flower clock based on our own biological day. It might not be as regular as Linnaeus'. Still we might be surprised to discover that besides hunger pains and bowel movements there are other physiological events that occur with some predictable regularity during each day. Are there times when you feel surges of energy and others when you are fatigued? When does your body seem to demand you stop whatever you're doing and go to bed? When do you usually awaken each morning? You might try taking your temperature every two hours day and night for several days and recording it. See if there is a regular pattern and if it has any obvious relationship to the activities that accompany each reading. Be careful not to forget that setting an alarm clock to wake yourself up every two hours for several days just may influence your body's rhythm.

Underlying individual measures of time there appears to be a rhythm common to almost all living things. Every 24 hours all living matter seems to go through regular changes. We have waking hours and sleeping hours, and our bodies even undergo regular temperature changes during the course of a day. We have our own biological rhythm or clock whose unit of about 24 hours is shared by everything that lives on earth. This rhythm is called *circadian* which means about (*circa*) a day (*di*+*an*).

What determines the length of a day for all of us is still not known. Some scientists believe that it is part of the information contained in all living cells, and others believe that it is determined by geophysical properties and activities of the earth and sun and moon that we have yet to discover. What is known is that the rhythm of our lives is measured by roughly 24-hour beats and so is that of all living things. However, within that overall setting of life rhythm there are many different ways in which particular changes are perceived and the flow from birth to death organized.

The bean plant—daytime

The bean plant—nighttime

People in different parts of the world do not organize the changes they experience in their lives in the same ways. For example, we have a seven-day week, and it is easy for us to fall into thinking that people

A Week -- Seven Days or Four Days?

who live with different weekly cycles seem strange or wrong or somehow unnatural. The way we learn to organize our experience becomes so familiar that we believe that time is given to us in nature. A week is naturally seven days, isn't it?

Not for all people. The Afikpo people of Nigeria have a four-day week which corresponds to the way they alternate buying and selling at the market with farming. Their week consists of these four days: *eke* is a nonfarming day when people from miles around go to a central market, *orie* is a farming day, *ahɔ* is another market day when people go to small local markets, and *nkwɔ* is a farming day.

This four-day week (with seven weeks in a month) seems as natural to the Afikpo as our seven-day week is to us. Which is the real week? Is there a real week? Or is time made by living creatures rather than something absolute and given in nature.

Different people perceive different changes in their lives and develop varied systems for organizing them. Yet these ways of organizing experience are nothing compared to the way time must be experienced and organized by creatures like a moth, a dog, an amoeba, a whale, a swallow, a housefly, a frog.

Focusing on Animal Time

There is an experiment in imagination that might help us move into these animals' worlds. Microscopes help us see things too small to perceive using the eye alone; telescopes help us bring things that are large but far away into our worlds. Imagine a time microscope and a time telescope. The time microscope would allow us to see changes that we ordinarily can't see: for example, every movement of a hummingbird's wing, every step taken by a cheetah in full pursuit of a deer, every step in a bee's dance. This instrument would stretch out time the way meditation sometimes does. It would fit into a minute of clock time a lifetime of experience. By stretching out a moment we miss things—things that move slowly and don't seem to be occurring at all. We might, for example, focus the time microscope on the bulb of a lamp. From the perspective of our eyes, the lamp gives out a steady light which we can read by. Focusing in on the bulb, however, we would lose that steady illumination—would see the flickering on and off of the filament which takes place too fast to be perceived ordinarily. Through the time microscope the bulb would seem like a strobe light and the room would seem to consist of shapes and forms that moved in jerks and jumps rather than of people and objects that had stability or moved smoothly and continuously. In fact, people would be difficult to perceive—they would be understood mostly in terms of what light did to them instead of as independently acting creatures.

If we focused the time microscope even more finely, the world of people and objects might dissolve altogether. We would be left experiencing the movement of neutrons and protons and electrons—a world in such constant motion that it might seem composed solely of the motion of small particles and empty space. Every second of our time would be filled with millions of perceivable changes—would provide a lifetime of experience.

Now switch instruments. Think about the time telescope which can see the growth of a redwood tree or the life and death of a star. In the perspective of stretched time, the life of a human being is like the one-night existence of a moth to one who can perceive the life and death of a star. Most likely a single life, a cosmic 1/100 of a second, would go unnoticed or seem a speck of dust on the telescope's lens.

Which is the real time? The only way to come to understand time as experienced by different people and different living beings is to give up the idea of one real time. It might help to think of those moments in your life when an hour seemed like a day, and a day passed so fast it seemed less than a minute. These feelings about time are not merely psychological—in your mind. The calendar and the clock are as much in the minds of people as the way they feel time. We need calendars and clocks to keep us from going crazy, to keep our lives together.

But we sleep every day, some days are longer than others, both in terms of the moments we experience and the way the sun moves. People need to fool themselves into believing the world is stable and regular. That's how we convince ourselves that we'll wake up to the same world tomorrow that we experience today. But we can imagine it differently and often feel it differently. Our personal experience of time is full of jumps and starts; of long days, longer days, and days that went so well and fast that we can hardly believe they happened. Yet our minds impose a smoothness upon time, a regularity that convinces us that everyone's time is the same, that time flows steadily and continuously no matter what our senses tell us. Our imaginations sometimes destroy the strongest convictions we have about reality.

What would happen if you woke up one morning and found out that you were a bug?

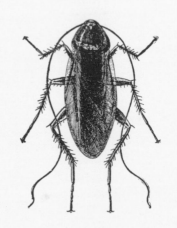

Cockroach (*Periplaneta americana*)

A Bug?

Franz Kafka begins his story *The Metamorphosis* with Gregor Samsa waking up to find himself changed into a giant cockroach or beetle:

As Gregor Samsa awoke one morning from uneasy dreams he found himself transformed in his bed into a gigantic insect. He was lying on his hard, as it were armor-plated, back and when he lifted his head a little he could see his domelike brown belly divided into stiff arched segments on top of which the bed quilt could hardly keep in position and was about to slide off completely. His numerous legs, which were pitifully thin compared to the rest of his bulk, waved helplessly before his eyes.

"What has happened to me?" he thought. It was no dream. His room, a regular human bedroom, only rather too small, lay quiet between the four familiar walls. . . .

His immediate intention was to get up quietly without being disturbed, to put on his clothes and above all eat his breakfast. . . . To get rid of the quilt was quite easy; he had only to inflate himself a little and it fell off by itself. But the next move was difficult, especially because he was so uncommonly broad. He would have needed arms and hands to hoist himself up; instead he had only the numerous little legs which never stopped waving in all directions and which he could not control in the least. When he tried to bend one of them it was the first to stretch itself straight; and did he succeed at last in making it do what he wanted, all the other legs meanwhile waved the more wildly in a high degree of unpleasant agitation. "But what's the use of lying idle in bed," said Gregor to himself.

He thought that he might get out of bed with the lower part, which he had not yet seen and of which he could form no clear conception, proved too difficult to move; it shifted so slowly; and when finally, almost wild with annoyance, he gathered his forces together and thrust out recklessly, he had miscalculated the direction and bumped heavily against the lower end of the bed, and the stinging pain he felt informed him that precisely this lower part of his body was at the moment probably the most sensitive.

Gregor Samsa's dilemma takes us out of our usual space and time. He tries to behave as he usually does, but his insect body and sense get in the way. Yet he is still a person who has become a bug. It is necessary to move even further away from thinking as a person to get closer to the way time exists in the lives of creatures who see and feel and move differently than we do. It is necessary to forget night and day, to think about time without thinking about weeks and years, without depending upon hours and minutes. Most of all, it is necessary to give up the idea that time flows regularly and steadily.

Imagine, for example, the dayless and nightless life of a blind fish living in a cave over a mile under water. Its time is determined by hunting, fleeing, resting, nesting, and mating rituals, and by changes in the movement of water. Light which is so central to our sense of time does not play a role in its life.

Think also of a moth that only lives a day and of turtles that live close to 100 years. Imagine the passage of time for hibernating animals like bears who sleep for months, or monarch butterflies who after flying several thousand miles from the eastern United States and Canada to a mountain 9,000 feet above sea level north of Mexico City spend the winter in a state of suspended animation, neither eating nor moving for months.

Time in the World of a Tick

The tick experiences time in a way that is even more difficult for us to understand. Most of its life is passed in a trancelike state of waiting, and its most important act is a form of suicide.

Jacob von Uexkull, one of the founders of ethology, described the life cycle of a tick in one of his works:

From the egg there issues forth a small animal, not yet fully developed, for it lacks a pair of legs and sex organs. In this state it is already capable of attacking cold-blooded animals, such as lizards, whom it waylays as it sits on the tip of a blade of grass. After shedding its skin several times, it acquires the missing organs, mates, and starts its hunt for warm-blooded animals.

After mating, the female climbs to the tip of a twig on some bush. There she clings at such a height that she can drop upon small mammals that may run under her, or be brushed off by larger animals.

The eyeless tick is directed to this watchtower by a general photo-sensitivity of her skin. The approaching prey is revealed to the blind and deaf highway woman by her sense of smell. The odor of butyric acid, that emanates from the skin glands of all mammals, acts on the tick as a signal to leave her watchtower and hurl herself downwards. If, in so doing, she lands on something warm—a fine sense of temperature betrays this to her—she has reached her prey, the warm-blooded creature. It only remains for her to find a hairless spot. There she burrows deep into the skin of her prey, and slowly pumps herself full of warm blood.

Common wood tick

Experiments with artificial membranes and fluids other than blood have proved that the tick lacks all sense of taste. Once the membrane is perforated, she will drink any fluid of the right temperature.

If after the stimulus of butyric acid has functioned, the tick falls upon something cold, she has missed her prey and must again climb to her watchtower.

The tick's abundant blood repast is also her last meal. Now there is nothing left for her to do but drop to earth, lay her eggs and die.

What is important is that the tick does not drop on any creature that happens to pass under it. It does not respond to rain or to the movement of the grass or to a branch dropping from a tree. The tick has an internal set—it waits for a particular smell, a particular temperature that is characteristic of mammals. It waits even though in most ways it is hardly alive. It is hard to imagine what the passage of time is like in such a passive yet strangely alert world.

Von Uexkull tries to get closer to the tick and guess at what its time might be like:

The tick hangs motionless on the tip of a branch in a forest clearing. Her position gives her the chance to drop on a passing mammal. Out of the whole environment, no stimulus affects her until a mammal approaches, whose blood she needs before she can bear her young. . . .

The lucky coincidence which brings a mammal under the twig on which the tick sits obviously occurs very rarely. Nor does the large number of ticks ambushed in the bushes balance this drawback sufficiently to ensure survival of the species. To heighten the probability of a prey coming her way, the tick's ability to live long without food must be added. And this faculty she possesses to an unusual degree. At the Zoological Institute in Rostock, ticks who had been starving for eighteen years have been kept alive. A tick can wait eighteen years. That is something which we humans cannot do. . . .

This ability to endure a never-changing world for eighteen years is beyond the realm of possibility. We shall therefore assume that during her period of waiting the tick is in a sleeplike state, of the sort that interrupts time for hours in our case, too. Only in the tick's world, time, instead of standing still for mere hours, stops for many years at a time, and does not begin to function again until the signal of butyric acid arouses her to renewed activity.

Think of the flow of time in the life of a tick. Time does not pass smoothly but is organized into periods of frantic activity and almost rocklike passivity. The time between a tick's taking up a perch and dropping onto another creature is but a moment no matter how long it might seem in our time. However, as soon as it hits the animal it scurries around looking for heat. At that stage of its life it senses many different things. The differences or changes it perceives with its senses give rise to its experience of time. The time of the tick moves unsteadily, at times frantically and at times hardly at all. The life cycle of the tick will not show steady movement from birth to death as it would for people. Rather it would show frantic activity at the beginning and end of life and slow changeless existence over the middle of its life. Using the image of the time microscope and telescope, one could say that compared to humans, ticks lived microscopic time at the beginning and end of their lives and telescopic time over the middle of its existence.

A Moment in Life

The fastest perceivable change that a living creature experiences can be looked on as a moment in its life. Anything that happens more quickly than that won't be perceived.

To understand this better try the following experiment. Take a pin or toothpick and tap it on your fingertip. You'll feel the tap. Then tap your fingertip twice quickly. You'll probably feel the two taps. Increase the speed of tapping. At a certain point you'll no longer be able to tell how many times you've been tapped, and if you can go fast enough, you'll simply feel one constant pressure rather than a series of taps.

Try the same experiment with some friends. Ask them to close their eyes and tell you how many times you are tapping them on the back of the neck or on the fingertips. If you begin fast enough, they'll only feel one tap. In fact, it has been discovered that 18 taps or more a second are felt as even pressure. Human tactile senses just don't register more than 18 separate impressions per second.

Our other senses also have limits. For example, the human ear can distinguish between 15 and 20 vibrations a second—depending upon the individual. As soon as the vibrations per second get over 20, we can no longer hear distinct sounds. If you could set up a drum that beats 25 times a second, you would hear one steady sound and not a very fast beat.

There are similar limits to what the human eye can see. Between 18 and 24 images a second are all that can be taken in as separate images. If things move faster than that, their images become blurred, and if the speed is constantly increased, there comes a time where even a dark, heavy object would be impossible to see speeding by our faces on a sunny day.

The limits of our eyes are what make movies possible. A film consists of a strip of still pictures projected on a screen so that we see movement rather than individual pictures moving jerkily before our eyes. The way that is done is by making the pictures pass before our eyes at approximately 1/24 of a second intervals. In that way, if the pictures differ just a little from one to the other, our eyes fuse the images. We see steady movement rather than single pictures.

It is even possible to use film to see movement that is too fast for our eyes. A camera can work faster and register changes we can't see. It is possible, for example, for a camera to take enough pictures per second to capture the movement of a hummingbird's wings. Then the pictures can be projected more slowly for us—at 1/24 of a second intervals.

An interval of between 1/15 and 1/24 of a second seems to be the fastest change in sound, touch, and vision we can experience. This interval varies slightly from person to person, and from sensory apparatus to sensory apparatus. However, it sets an approximate limit to the way we perceive change in the world. For that reason somewhere between 1/15 and 1/24 of a second could be described as an approximate moment of human experience. Moments give you an indication of how creatures usually experience the flow of time. They define what is the normal pace of life in a creature's world.

A Moment in a Snail's Life

For us the snail creeps along, but to understand the way a snail experiences things and have insight into the normal pace of a snail's life, we have to learn something about what the limits of its senses are, and what a moment might be like for it.

There are some experiments that have been done to determine what a moment is for the snail. In one of them a snail is placed on a rubber ball which is floating in water so the ball can move easily. The snail's shell is held in place by a bracket so that the snail can crawl on the ball. The snail can make its normal movements even though it's held in place, since the ball moves in the water. If a small stick is moved next to the snail's foot, the snail will climb up on it. If the snail is given one to three taps per second with the stick, it moves away from it. But if four or more taps per second are given, the snail will climb onto the stick. In the snail's world the stick which moves back and forth four or more times per second becomes stationary. Three or four moves per second seem to be what the snail's sense perceives; $1/4$ of a second is an approximate moment for the snail.

Thus, a lot of things that seem full of movement and change to us seem unchanged to the snail. Imagine experiencing the world that way —grass you see moving would be still, cars speeding by would not be part of the moving world. Four taps per second on your fingertip or the back of your neck would be felt as constant pressure. The snail, fortunately, lives well with its pace; what it can't perceive is most often quite irrelevant to its search for food and a place to deposit its eggs.

However, there are limitations—the movements of a bird toward a snail wouldn't be perceived. The snail would not feel the vibrations set up by the bird's wings and would be in the bird's beak without any warning. Our senses can protect us and also can leave us defenseless. Think, for example, of all the pollutants that we cannot detect with our senses until they have already caused disease in our bodies and damage to the environment.

The Vibrating Web

Many animals perceive things that are without motion or change for us. Vitus B. Dröscher describes the vibrations spiders can perceive in his book *The Magic of the Senses*. He says that for a spider:

the web serves as a telegraph line by which various messages can be perceived. All web spiders live in a tactile world. Their visual sense is little developed and plays a very minor part in their lives. If an opaque mask is stuck over their eight eyes, they live, mate, feed, and flee just as well as sighted spiders. But how do they know whether it is prey that has fallen into their toils or an enemy, whether a suitor is approaching or a young spider trying to make off?

These questions have been investigated by Dr. Erwin Tretzel of the University of Erlangen whose work is described in *The Magic of the Senses:*

The male of the garden spider "phones" his chosen bride. He attaches a thread to her web, which he plucks in a certain rhythm. Related species cling to such a thread in order to make it vibrate by jerky movements. By the way the female reacts and by what kind of vibrations she sends out in her turn, the male knows how great the danger is of being eaten by his bride, or whether he can risk mating. . . . The spider mother has a standard-

ized warning signal. . . . While the movements used to call her young make the web vibrate slightly and softly, warning is given by a short, violent jolt to the web caused by rapid movement of a hind leg. This warning serves to send forward youngsters, who want to accompany their mother in her assault on the prey, back to the safe hiding place. The mother stops briefly, moves one hind leg, and it is amusing to watch how the obedient youngster immediately turns tail and hurries back. The young are also sent back as long as a captive prey is still struggling hard.

To perceive these vibrations which are so crucial to the spider's existence, the moment of a spider must be such that distinct vibrations are experienced. If the senses of a spider were like those of a snail, the web would seem still. What is a moment in a snail's world is full of change and variation in the world of a spider.

It is possible to get closer to the world of a spider than with most animals. Find a web that is intact. Then slowly and quietly explore the web, looking in corners or on twigs until you find the spider in its home. Wait as a spider does and observe what happens when a fly or some other insect gets trapped. Watch how the spider sits and orients itself. How can it feel the vibrations in the web? Does it smell or see its food or does all the information come from the web?

If you are lucky enough to see a small sack of eggs woven by a spider, watch it ten minutes in the morning and ten in the evening every day. Train yourself to be patient, to observe animals over a long period of time, to wait and not be frustrated if from your perspective nothing seems to be happening. The worlds of animals unfold at different paces than ours—some slower and some so quickly that they escape us. But over a period of time it is possible to catch hold of some of the differences and feel something of the beauty and variety of different forms of life.

The Moment
and Communication
among Animals

Communication takes place over time. We talk, others listen and respond. If we talk too quickly, we can't distinguish sounds or words and can't understand each other. The same thing happens when we speak too slowly. Put a 33-rpm record on at 78 rpm and a 78-rpm record on at 33 rpm. They will give you an idea of the pace a series of sounds must have to be distinguished as language by us.

People can distinguish sounds that range from about 15 to 15,000 hertz (i.e., *cycles* or vibrations per second). The dolphin can distinguish sounds between 400 and 200,000 hertz. Its language, which is quite complex, is too fast for us to distinguish individual sounds, and its voice is high pitched. Using a tape recorder it is possible to slow down a recorded dolphin conversation and begin to hear individual sounds, though the language is still not understood well enough for us to communicate with dolphins.

Not all communication is done through sound. Bees communicate through touch, smell, and dance. By vibrating their antennae two bees can identify each other. Furthermore, every bee has a scent organ called a Nassonoff's gland which releases particular scents when a bee bends its body. Each hive has its own scent and each flower has its own scent. By varying the scents it emits, a bee can inform other bees which hive it belongs to and which flowers it visits. Imagine being able to smell so finely and being able to translate almost simultaneously the scents into useful information. It is as if you could control the smells emitted by your body and use them to indicate who your family is and what food bargains you found in the supermarket.

Time, the moment, is crucial in all these forms of communication which depend upon small impulses of sound or scent or gesture or vibration being varied and distinguished. Without the ability to distinguish in time a succession of impulses, no communication is possible.

The ability to perceive small changes and get information from them is probably most dramatically illustrated by the dance of bees. When a bee discovers a field of flowers, it returns to its hive and begins a dance. The dance begins inside the hive with a fluttering movement. The bee discharges a scent indicating the kind of flower it has discovered. Then it takes flight, tracing a figure 8, raising the tip of its abdomen, and fluttering its wings. Karl von Frisch, a biologist who first uncovered the language of the bees, calls this a round dance. The center line in the figure 8 (see the diagram) indicates to the other bees the direction of the flowers compared to the direction of the sun at that time of day. The speed of the dance indicates how many flowers there are and how good the nectar is.

After this complex dance, which to our eyes would merely seem the random buzzing of a bee, the messenger flies outside the hive and begins a tail-wagging dance. This indicates the specific direction of the flowers and the distance they were from the hive. This dance provides information that is accurate to within inches. Von Frisch once moved a dish of sugar water a bee discovered and described to other members of the hive about a foot from where it was discovered. The bees, however, headed straight for the spot the sugar water was discovered. The information they received was completely accurate.

The whole performance lasts no more than half a minute our time. Imagine what it would be like to convey all of that information in gesture, mime, and dance in 30 seconds. Movements of a bee's antennae and body which seem continuous to us are actually jerky, composed of smaller and larger gestures, of pauses and accentuations.

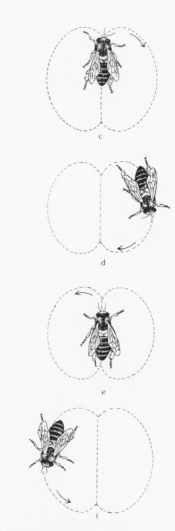

The wagging dance of the honey bee

The closest we can get to feeling that visual moment and the language it makes possible is to observe the sign language of the deaf. Sign language involves facial expressions, hand movements, body position. It is full of ideas and concepts we can hardly express in oral language and requires a quickness of eye few of us develop. For example, the sign for "little" is made by holding the tips of the thumb and index finger apart with the rest of the fist closed. If the tips of the thumb and index finger are about an inch and a half apart, that indicates something is small but not minute. As the tips of the fingers move closer together, the concept of small gradually transforms itself into other degrees of smallness such as tiny, minute, microscopic, and to such psychological concepts as insignificant, lightweight, selfish, and miserly as well. A whole continuum of size can be represented in a way that is almost impossible to express in words. Not only that, the hand can move so fast that ideas can follow each other with a speed that is not possible with words. Moreover, with two hands moving two ideas can be expressed simultaneously. It would be very interesting to study the moment in the umwelt of deaf individuals. It may turn out that it is possible to cut time more finely than we usually do and perceive change in finer and more sensitive ways than we are accustomed to.

The Creation of Time

All animals don't respond in the same ways to things that happen in the world. The senses are built differently and select different aspects of things. Some birds have an interesting and not fully understood ability to perceive very slight vibrations. Robins, for example, can sleep in trees shaken and vibrated by the wind. They act as if they don't perceive the vibrations at all, yet they awake with a start and fly off

after perceiving the slight regular vibrations caused by a marten climbing the tree to hunt them. Time is very complex in such lives which
perceive very fast changes and some too slow for our human senses to
pick up.

There is a dramatic example of how animals perceive changes even
the finest scientific instruments can't. In the People's Republic of
China a number of scientists noticed that many farmers could predict
when an earthquake was about to take place. When asked how they
did it, they explained that the birds and horses always acted strangely
several days before the first humanly perceivable tremors of the earth.
They concluded, and there is no reason to doubt it, that these animals
were perceiving movements in the earth that people even with their
most sophisticated instruments couldn't detect.

Bees also perceive tremors which are too slight to be of concern to
people. When bees fly from their hives to flowers they have to find
their way home. However, the wind constantly throws the bee off its
path. What is a windless day for us is full of gusts and drafts and currents for a bee. The bee's senses have to be able to distinguish these
slight movements of air. What wouldn't be perceived as movement to
us has to be organized in a bee's world or it would never be able to
navigate back to its hive. There are approximately 2,500 hairs in a bee's
eye which measure each gust of the wind. They are arranged so that
they do not block the bee's view but they respond like levers to any
gust of wind. Somehow the bee is able to organize the information it
receives about these minute changes in wind so that it can change its
flight maneuvers and keep on a steady path to the hive.

One of the questions that natural scientists and biologists still
have to answer is how this information is used by the bee. We can understand how finely the bee cuts experience, how much experience

exists in a bee's world which is too fine for us to perceive. But the question of will, choice, feeling, and integration, of what the total experience of a bee is like, is open and may not even be answerable.

A moment in a bee's life does not even approach the limits of what small changes some animal senses can perceive. There is a South American knifefish that can emit and distinguish as many as 1,600 electrical impulses a second.

From the knifefish to the snail, time is experienced in different ways. Yet to us humans the question keeps coming up—what in all this is the real time? How does time really flow? Isn't there a steady movement from past through present to the future, no matter how different creatures organize and exist within that flow? The answer seems to be that time exists as it is experienced and organized; that there is no absolute time, only lived time. Each animal creates its own time, and the best we can do is try to understand how its temporal moments relate to its behavior.

PART IV

Tone, Mood, and Response

Tone, Mood, and Response

Whenever I find myself growing grim about the mouth; whenever it is a damp, drizzly November in my soul; whenever I find myself involuntarily pausing before coffin warehouses, and bringing up the rear of every funeral I meet; and especially whenever my hypos get such an upper hand of me, that it requires a strong moral principle to prevent me from deliberately stepping into the street, and methodically knocking people's hats off—then, I account it high time to get to sea as soon as I can.

Herman Melville
Moby Dick

Try this experiment on yourself and some friends. Make up a deck of color cards consisting of an intense and dull shade of the following colors: yellow, blue, violet, green, and red. Also add a black and a white card. You can use markers or crayons to make the cards. The deck can also be made out of paint sample cards that are available free at most hardware and paint stores.

Mix up the deck and place the cards color up on the table. Then arrange the cards in order from those you like most to those you like the least. Write down your choice and have other people do the same. Do all the people make the same order? What causes the differences?

Usually there is a large variation in people's color preferences. Not only that, choices differ from day to day. Try to order the color cards on a day that you're feeling happy and on a day when you're gloomy or angry or tired or frustrated. Color preferences change with moods. This is reflected in our language. We talk about feeling blue, seeing red, being in a black mood, or in the pink. People go green with envy or purple with rage.

Our moods determine the way we experience and respond to things. If we feel weak or scared or nervous, ordinary things seem full of menace. Shadows and noises represent possibilities of attack; a casual glance can seem a challenge. The same food can seem tempting and attractive when we're hungry and inedible and unappetizing when we're full.

The mood of a person determines the aspect of experience that becomes important. People in love don't pay much attention to food. Enemies look for each others' weaknesses where friends pay more attention to strengths. Someone who is outraged can overlook kindness, and loneliness often leads to a supersensitivity to any sign of communication.

Looking for a Home

People are not the only moody animals. Friendly dogs greet each other by wagging their tails while enemies growl and bare their fangs. A threatening angry cat humps its back and hisses while a contented one purrs. In order to understand how mood functions in the lives of other animals, it is necessary to understand how they experience the world. Take the hermit crab for example. This small animal is not a true crab or a lobster or shrimp. The naked hermit crab, out of the shell it uses for a home, is a curious-looking creature. From the front it looks a bit like a crayfish. Its eyes are mounted on the ends of stalks which move all around pointing the eyes towards objects the way we point a flashlight. It has two pinchers, one large and the other small. The hermit crab is not a handsome symmetric creature. Not only is the right claw larger than the left, its tail curls to the right so strongly that it almost looks as if there should be another left tail to balance it. Between the tail and the claws are six legs, three on each side. There are two feelers coming out of its head near a tight, pointed mouth. These feelers are large, and they bend down along the path the hermit crab travels almost as if it were a blind creature using two canes to feel its way through the world.

A close look at the hermit crab shows it teeming with hairs on its legs and feelers and body. These hairs are like dog's or cat's whiskers. The hermit crab feels its way through the world, sometimes on the sand, sometimes under a rock or in the water swept by the movement of the tides. The crab's eyes indicate shapes and forms but are weak compared to the sensitivity of its hairs. Imagine feeling your way through the world with every hair on your body. All the hairs on your head would be sensitive organs of touch. Surfaces that ordinarily seem smooth to us would feel rough and irregular. No two pieces of paper or wood would feel the same to us.

Even with such sensitivity the hermit crab is a vulnerable animal. Its tail and body are soft, it doesn't sting or have poison pinchers, it doesn't dart or respond quickly. The main ways it protects itself are

by finding a hard mobile home that it carries around and by helping poisonous creatures grow on this shell home and share its food.

The hermit crab without a home searches for an empty shell by feeling its way under rocks and through tide pools. In its world it is the weight, shape, and feel of a shell that is crucial. Once it finds an empty shell, its curved tail comes into use as it snuggles inside the shell and fills it up. Snail shells seem to be particularly well suited for hermit crabs. They spiral around and the asymmetric hermit crab curls right in and makes a door using its large claw. It can seal itself in its home and shut the door. It can also keep itself anchored in the shell with its tail while moving, house and all, in search of food. This still leaves the hermit crab quite vulnerable to fish, starfish, gulls, and other predators.

Some hermit crabs have developed a relationship with a group of anemones that benefits both creatures. Relations like these are called symbiotic relations. Once it has found a shell, the hermit crab finds a special kind of anemone that has tentacles which sting and irritate fish and other potential enemies. In some cases the anemone climbs on the hermit crab's shell and plants itself there. In other cases there seems to be some complex communication based on touch between the hermit crab and the anemone. They tap and touch each other, and the anemone then releases its grip on the rock it clings to and plants itself on the hermit crab's shell. Then they move off in community, the hermit crab hunting and sharing its food with the anemone, and the anemone protecting the whole little community.

However, things don't always work smoothly between the anemone and the hermit crab. That's where mood comes in. If the hermit crab has a shell but has no anemones on the shell, it tries to plant the anemones on its shell. If it has no shell, the anemone takes on a whole different character. The hermit crab approaches the same creature, only this time it is in a mood to find a home. The anemone is a potential dwelling rather than a potential symbiotic partner. The hermit crab tries to climb inside the anemone and use it for a shell.

There is a third case where the hermit crab has a shell and a colony of anemones on its back but has not been able to find food for a while and is starving. Then an anemone that it might implant, crawl inside of, or leave alone on other occasions, becomes a source of food and is attacked. It is the same anemone in each case, the same subject in the hermit crab's umwelt. But the subject is treated differently according to the mood or disposition or needs of the hermit crab. The animal does not act the same way every time it faces the subject.

If you decide to observe animals in their natural environments, it is crucial to understand that in the world of all but the simplest of creatures there is a range of responses. In order to understand animals, it is necessary not only to understand the world as they experience it, but to be more specific and understand how an animal feels at the moment that you're observing it.

The Cats Will Play

Cats chase mice and eat them—but not always. Some cats become attached to objects they play with such as balls of yarn or rolled up paper. They practice hunting, pounce on the yarn or paper, and toy with it as they do with mice. These objects have a special revered place in cats' lives, much like a child's favorite stuffed animal or toy. Paul Leyhausen who has spent years studying cat behavior describes a male fishing cat's unusual attachment to a large yellow plastic ball. The cat played with the ball all the time and it was covered with teeth and claw marks. Once Leyhausen put a female fishing cat in the male's cage.

The male sat quietly in the corner and allowed the female to examine everything in the cage and walk about freely. However, when she approached the yellow ball and sniffed it, he pounced on her and bit her furiously on the neck. Leyhausen had to separate them or the female would have been seriously injured.

However attached cats become to toys and play at hunting and catching them, they don't mistake these toys for real mice. They feel and smell and see the difference. A hungry cat goes after a live mouse. On the other hand, many cats when they're not hungry prefer their toys to live mice. A full and playful cat will often ignore a mouse that walks right past its nose.

If you have a pet, watch it when it wakes up; when it's ready to go to sleep; when it hasn't eaten for a time; after it's had a good meal; when it seems restless and wants to move and play. An interesting experiment is to offer your pet food at all these times and see if the food is refused. Another experiment is to offer it a toy at all these different times, and see what happens. You can also observe how it responds to people when it's hungry or sleepy or exhausted or full of energy and wants attention or exercise. We've noticed that Sandy, who's the gentlest of dogs, will growl at anyone if he's hungry and has a bone. He also goes crazy and chases his tail jumping and running in circles if he sees us pick up the car keys or his leash. He gets jumpy on rainy days when he sleeps inside and doesn't move around. Bit by bit we're piecing together a picture of his moods and needs.

Recognition

We recognize people primarily by the way they look and the way they sound. From a distance it is hard to tell people apart if they're

The herring gull chick responds
to the stick, not to the model of
its parent's head.

standing still. If they're moving we get more clues. Some people are easy to recognize by their gestures, by the way they swing their hips or move their arms. As soon as we can make out someone's face or hear their voice, recognition is no longer a problem. We not only can tell friends from strangers, we can also guess at people's moods from the way they look and sound. For animals who do not depend upon their ears and eyes as we do, recognition takes place in many different ways. Yet within the space and time of most animals other members of the same species are clearly distinguished. Just as we depend upon faces and voices, animals depend upon certain details to recognize other members of their own species. These details are called key stimuli.

It has been discovered, for example, that a herring gull chick is attuned to its parent's beak which is red with three white rings. Niko Tinbergen discovered that these gull chicks respond more to a thin red stick with three white rings than to a realistic though uncolored plaster head model of a herring gull adult.

Our son Joshua inadvertently discovered a key stimulus for dogs. One day he was playing Dracula. He had some plastic vampire teeth in his mouth. These teeth made Josh look as if he were baring his fangs. He playfully approached a neighbor's dog who is usually friendly and imperturbable. The dog took a glance at the fangs and started to growl. He snarled and bared his fangs back at Josh who was by then so terrified that he let the teeth drop out of his mouth. As soon as that happened, the dog relaxed and became his friendly self again. He had reacted to Josh's fangs just as he would have to those of any menacing dog.

Ethologists set up many experiments to discover the aspect of experience that causes recognition. These experiments try to isolate the key stimuli from all the other aspects of the animal's experience. Here are some examples. Male toads approach all moving objects during the breeding season and clasp onto them. The male lets go only if the clasped object gives the cry of another male toad. Otherwise, the toad will hang on, even if it is clasping a fish or a human finger. Young black-birds respond to a simple two-headed black cardboard model the way they do to their parents. Young stickleback fish respond to wax models that are colored like their mothers regardless of the size or shape of these models. They don't have to look anything like fish—color is the key stimulus.

This isn't so strange. How could we tell the difference between a wax museum dummy and a real person, or a cotton dummy seen from behind and a real person seen from the same view? There are certain situations in which we couldn't tell.

Finding and Understanding the Keys

We can get closer to the worlds of animals by understanding the key aspects of experience they respond to. It is possible to experiment with our language and try to portray the ways that animals use certain keys to identify other members of their own species. Here are three short tales of recognition. Try to follow the language and not imagine the action in visual form. Flow with the images and the senses they are drawn from. Recreate the rhythm and speed that the description indicates. And remember they are still only guesses.

---

1. A Monologue

Windy go to light go to light   that scent forget light that scent only that one eliminate light concentrate scent move on to scent windy scent come scent go scent come scent go follow rhythm of scent wind correct back to scent to scent to scent scent SCent SCEnt SCENT—touch

---

This monologue is a scent story narrated by a male moth. Female moths emit a perfume too fine for our noses to distinguish. They can separate this scent from all the others in their environment and detect it up to a mile and a half away. The scent carries an important message. It indicates the presence of a mate and, as in the narrative, causes all other activity to cease. The male moth sets out on a journey following the scent trail which is affected by the wind and by other scents. If he is lucky, he can follow it as it increases in intensity until he reaches the source of the scent and touches his mate.

---

2. *A Speedy Dialogue*

First player: anyonearound anyonearound anyonearound anyonearound

Second player: I'mabout I'mabout I'mabout I'mabout I'mabout

First player: whowhereareyou whowhereareyou whowhereareyou whowhereareyou

Second player: femaleoneofyourkindoverhere femaleoneofyourkindoverhere

First player: I'monmyway I'monmyway

The dialogue is between two dolphins about 100 miles apart. They can talk at this great distance under water. One dolphin sends out a message, the other answers, and they carry on a conversation. Scientists have discovered that dolphins can communicate on a complex level. They send out messages at least 16 times more rapidly than we do. To get an idea of this put a 33-rpm record on at 78 rpm. The speed of the 78 makes the sounds high pitched as well as speedy. Now imagine putting a 33-rpm record on at a speed of 528 rotations per minute. That is closer to how the dolphin talks. As in the dialogue, the messages are complex and produce recognition as well as enough information for the dolphins to rendezvous if they choose.

---

3. *A Conversation on CB Radio (at 300 pulses a second)*

| SENDER ONE: whoareyouwhoareyou | uoyeraohwuoyeraohw :SENDER TWO |
| SENDER ONE: whoareyouwhoareyouwhoareyohwhoyeryohwuoyeraohw :SENDER TWO |
| SENDER ONE: | :SENDER TWO |
| SENDER ONE: *whoareyouwhoareyou* | *UOYERAOHWUOYEROHW* :SENDER TWO |

This dialogue represents perhaps the most unusual of the three cases of recognition. A fish called *Gymnarchus niloticus* has a similar communications system that consists of an electric organ which continuously emits 300 pulses per second of a 3- to 10-volt electrical current. This current sets up a magnetic field around the fish with its head being the positive pole and its tail the negative. The pulse beat sets up a wave movement in the water that is quite similar to radio waves in the air. When two *Gymnarchii* are in the same area, these waves interfere with each other, disturbing the fish's capacity to use their electric sense to navigate or search for food.

The situation is similar to what happens when two people try to broadcast on the same wavelength at the same time. However, the *Gymnarchii* have a way of dealing with the problem. As soon as some interference develops, both fish stop broadcasting, then they readjust their frequency rates so that each one broadcasts on a different frequency. Then they can distinguish each other's sounds while broadcasting simultaneously. There is no longer any interference, and they can use their electric sense once again.

Friendly Enemies

Animals recognize their own kind, their food, and their enemies. The way recognition takes place is an umwelt problem. It is determined by the way the senses of the animal function and by the space and time it lives in. The umwelt frames the lives of animals. Within this frame, however, many complex activities unfold. Rattlesnakes, for example, live within a space defined by temperature variations, as mentioned earlier. Male rattlesnakes have a defined territory and can tell when that territory is invaded by another male rattlesnake. There is a temperature specific to rattlesnakes as well as to their prey. Once a male rattler identifies the presence of another male rattler, an interesting thing happens. It does not assume a hunting mode whether it is hungry or not. It becomes aggressive, the mood is one of defense of territory. The two males engage each other in a battle. They intertwine bodies, pound each other with their heads, and wrestle until one gives up from fatigue and leaves the territory to the winner. No matter how fierce the battle, they never bite each other. Their poison is reserved for prey, not rivals.

The oryx antelope acts in a similar way. The oryx males have long extremely sharp horns. They stab attacking lions with these horns and have been known to kill them. When two male oryx fight, however, they never use their horns to stab. They engage in a head-to-head pushing contest, and if their horns happen to touch or graze each other's body, they pause in their battle and start over again.

The lives of animals are determined by complex moods. There is behavior appropriate toward mates, families, enemies, rivals, predators, prey. There is behavior that happens when they're hungry that doesn't occur when they're full. There is mating behavior, protective behavior, gentle and affectionate behavior. The closer one looks at animal life, the more thoughtful and conscious animals seem. Yet the questions of whether animals think, make choices, and are aware of themselves and others have not been answered. There are intriguing hints about some possible answers though—worth exploring.

In India rhesus monkeys have lived in cities for centuries. They are not pets and roam loose in bands that can have as many as 70 members. Usually there are a few adult males, about three times as many adult females, and a number of babies and young monkeys. They live on rooftops or in abandoned buildings and are quite different from their forest cousins. In adapting to urban life they have changed their tastes and habits in ways that make one wonder whether conscious adaptation took place at some time.

First of all, they live in fixed dwellings and return home to sleep in the same place every night. In the forest the band wanders much more and has no fixed home, but settles in a different tree each night.

Second, city monkeys live in a continuing and tense relationship with people. They prefer cooked food to the raw food that is their preference in the forest. They don't cook, however, but forage. They raid houses, food shops, and markets. Sometimes they even snatch food

City Monkeys

from people, particularly children, and injure them in the process. These monkeys are often chased and beaten by shopkeepers and angry parents. Yet according to Sheu Don Singh, a psychologist who has studied them, city monkeys prefer living in cities to the forest. If they are captured and brought to the forest, they hurry back to the city as soon as they are released.

Urban monkey bands stay together all year, and the monkeys in a band take care of each other. Singh observed a baby monkey fall into a well. According to him, immediately most of the members of the group it belonged to rushed up to the well chattering. A few were ready to go down into the well until some people came along and pulled the baby out.

Imagine the fugitive life of an urban monkey. Threat and defense are major characteristics of its umwelt. The overriding mood of a life where food has to be stolen from others and where attack is always possible is aggression. Urban monkeys fight with each other over territory and homes, fight with people over food, and fight for social position within their own groups. The city monkey must be shrewd and alert all the time, much like its human counterpart in the city.

When city monkeys are put together with forest monkeys, they are indeed much more aggressive. City female monkeys establish dominance over forest males, a very unusual situation in the monkey world where males are usually the dominant pack leaders.

Forest monkeys are shy and flee when people come near them. City monkeys aren't afraid of people at all. They'll come close to people, snatch food, fight back when menaced. The aggressive mood of their lives and their compulsive attraction to the city reminds one of people. Many people move to urban centers, adapt themselves to the aggressive fast tone of city life and find they can't return to a quieter, slower country style. What is gained for both monkeys and people in aggressiveness and shrewdness is paid for by the loss of a calmer and simpler life. Do monkeys and people consciously choose to change? Do changes take place so fundamentally within their umwelts that once in a city world their horizons change, their moments quicken, their spaces constrict, and their tones intensify? If this is the case, if the umwelt changes so fundamentally, then it is no wonder that the city is preferred to the country. The umwelt of a city dweller, whether monkey or person, doesn't fit the country environment. If the umwelt doesn't fit the environment, the animal has to change, die, or leave. The dolphin can't communicate out of water, the cat can't live in water; yet at some time in the past dolphins were land creatures and cats' ancestors were water creatures. It's possible that one major cause of evolution and change in the animal world is the lack of fit between the umwelt and the environment some animals experience.

Buridan's Ass

Golden retrievers like Sandy are very gentle dogs. Our children can jump on him, pull his tail, put their hands in his mouth. He takes it all good naturedly and often seems to enjoy it. The most he ever does to object to their sometimes aggressive attention is get up and walk away. If he has a bone though and is hungry, he'll growl if anyone approaches.

Still there are times when not even bones dripping with meat and smelling of blood attract him. One day we gave Sandy a lamb bone. He was sitting on the doorstep with his favorite stick right in front of his nose. We put the bone down next to the stick and he stood up. His nostrils expanded, he stiffened his body, then smelled the stick, the bone, the stick, the bone, and looked at us with a confused and quizzical expression. At that moment he seemed a person. He was faced with a decision, torn between two strong instincts. It may be that situations like this lead to the development of consciousness—it's hard to know. Sandy either had to make a choice or stand facing the stick and the bone until he dropped from fatigue.

Sandy's confusion which left him facing us and looking a bit crazed was not unique to him. A French philosopher, Jean Buridan who lived during the 14th century, posed a problem that represents Sandy's dilemma and has come to be known as the problem of Buridan's ass. There was a very hungry ass. The animal had been working all day and was finally led to the stable where he was usually fed. He was led to two bales of hay, both identical, and was placed exactly between them. Which stack would he choose to eat from? They were identical and equally distant from him. According to one version of the story he couldn't make a choice and starved to death.

The problem of Buridan's ass is the problem of freedom and choice. Given two equally attractive and possible alternatives how does one make a choice? For Sandy there were two powerful instincts conflicting —hunger and retrieving. There are similar choices for people. How do you choose between two dishes on a menu both of which are your favorites? How do you choose who to fall in love with if there are many people you find equally attractive? How do you decide upon what work to do if many attractive possibilities exist in your life? As we suggested before, it is possible that consciousness, an awareness of yourself as different from others, and the dawning of the ability to control oneself may grow out of the conflict of equally strong instincts.

There are indications that choice as well as the development of awareness and intelligence are problems on almost all levels of animal life. Konrad Lorenz, one of the founders and most creative of ethologists, describes a moment of challenge and possible growth in the life of a cichlid, a radiant jewel fish. Male cichlids care for their young. If a baby strays from the nest, the father swims to it gently inhaling the fish into its mouth and returning it to the nest where it contracts its muscles and becomes heavier than water, therefore resting in the nest. According to Lorenz:

I once saw a jewel fish, during such an evening transport of strayed children, perform a deed which absolutely astonished me. I came, late one evening, into the laboratory. It was already dusk and I wished hurriedly to feed a few fishes which had not received anything to eat that day; amongst them was a pair of jewel fishes who were tending their young. As I approached the container, I saw that most of the young were already in the nesting hollow over which the mother was hovering. She refused to come for the food when I threw pieces of earthworm into the tank. The father however, who, in great excitement, was dashing backwards and forwards searching for truants, allowed himself to be diverted from his duty by a nice hind-end of earthworm (for some unknown reason this end is preferred by all worm-eaters to the front one). He swam up and seized the worm, but, owing to its size, was unable to swallow it. As he was in the act of chewing this mouthful, he saw a baby fish swimming by itself across the tank; he started as though stung, raced after the baby and took it into his already filled mouth. It was a thrilling moment. The fish had in its mouth two different things of which one must go into the stomach and the other into the nest. What would he do? I must confess that, at that moment, I would not have given twopence for the life of that tiny jewel fish. But wonderful what really happened! The fish stood stock still with full cheeks, but did not chew. If ever I have seen a fish think, it was in that moment! What a truly remarkable thing that a fish can find itself in a genuine conflicting situation and, in this case, behave exactly as a human being would; that is to say, it stops, blocked in all directions, and can go neither forward nor backward. For many seconds the father jewel fish stood riveted and one could almost see how his feelings were working. Then he solved the conflict in a way for which one was bound to feel admiration: he spat out the whole contents of his mouth: the worm fell to the bottom, and the little jewel fish, becoming heavy in the way described above, did the same. Then the father turned resolutely to the worm and ate it up, without haste but all the time with one eye on the child which "obediently" lay on the bottom beneath him. When he finished, he inhaled the baby and carried it home to its mother.

PART V

Views of the Oak

Views of the Oak

Erica's sweater: it is blue, has long sleeves, buttons down the front, is finely woven, smells of wool; it is formless, of no use but to snuggle up to and sniff, it smells like Erica, is good to stay near when no people are around; it is gigantic, has strands, is delicious, the perfect nest for eggs, a warm and nourishing environment to grow up in.

Which is the real object—the woolen sweater Erica our daughter wears? The formless object that Sandy takes from Erica's bedroom and rests his nose on when we are gone? The home and feast for the moths we found in it the other day?

An object or living being appears and functions differently in different umwelts. A dog is not the same to its fleas as it is to its owner or to another dog who is trying to size it up. There are many different

Consider the Oak Tree

perspectives from which things in the world are experienced. Consider an oak tree and all the life it can harbor from its roots to its highest branches.

Imagine first a fox whose den has been dug out among the roots of the tree. From the perspective of the fox the tree is home. It protects his mate and their offspring, and is a place to sleep and rest and eat. Because the fox is so vulnerable if trapped in the den, its entrance has to be hidden and appear in no way special. The fox has to behave as if that particular oak is nothing special and yet know every smell and sound surrounding it, as well as all the tunnels and dead-ends leading through the roots. The roots themselves are the columns and beams of the foxes' home which is lined with leaves to warm and hide young foxes.

The fox's view of the oak is low down, under and on the ground. There is no need in his or her world to look up. The trunk leads somewhere else, to other worlds and other experiences—it is neither threatening nor comforting in the fox's world. It is a part of experience that is background. The branches of that oak and of all the other trees that surround it do not separate themselves. There is a lattice of branches above the fox's world, yet those branches are phantoms, horizon features that have no connection to the oak in the umwelt of the fox.

Somewhere within that lattice there is a good chance that an old and abandoned crow's nest exists. And there is also a good chance that it has become home for an owl. Many owls are extraordinary hunters and lazy builders. The owl in our imagined oak is one of these and has taken up residence within the upper branches of the oak in the crow's nest. Most likely, the owl at the top and the fox in the roots will never cross paths. In some ways it doesn't even make sense to say they share the oak, since there is nothing shared about their experience of the tree.

During the day the owl and the tree are one. The owl sleeps, its feathers drawn into itself, resting so still that to human eyes it appears to be another branch. The color of its feathers helps. During the day it blends with the tree. At night the owl comes alive and the tree is a tower from which its keen eyes survey its hunting grounds. Owls' eyes are immobile—they don't move around, though they can focus on particular objects with amazing concentration and magnitude. The owl's eyes are like telescopes mounted on the swivel stand of its head which can turn 180 degrees in either direction. The eyes don't move but the head does. At night, for a while the owl sits in the oak, its eyes wide open, scanning 360 degrees. As soon as it spots prey, it is gone but usually comes back to eat. If the oak provides a protective environment during the day, it becomes a dining room through the night. Single owls have been known to return to their repossessed crow's nest with as much as a dozen mice, several rabbits, a mole, and ground-hog or two during three or four hours of hunting.

Halfway down the oak on and under its bark other worlds exist. Jacob von Uexkull describes the world of the bark-boring beetle and the ichneumon fly in his essay. *A Stroll Through the World of Animals and Men* which originally gave us the idea for this book:

The bark-boring beetle seeks its nourishment underneath the bark which it blasts off. Here it lays its eggs. Its larvae bore their passages underneath the bark. Here, safe from the perils of the outside world, they gnaw themselves farther into their food. But they are not entirely protected. For not only are they persecuted by the woodpecker, which splits off the bark with powerful thrusts of its beak; an ichneumon fly, whose fine ovipositor penetrates through the oakwood (hard in all other umwelts) as if it were butter, destroys them by injecting its eggs into the larvae. Larvae slip out of the ichneumon eggs and feed on the flesh of their victims.

And what about the oak tree in the umwelt of people? It is probably characteristic of our species that the human umwelt differs from person to person and culture to culture, and that it changes over time as well.

The oak tree has often had a special place in the umwelt of some peoples. To many Native Americans the oak is honored as a symbol of strength and longevity, and as a source of food. Acorns are washed, pounded into meal for bread, added to soup. In Europe, centuries ago, the druids believed that the oak harbored spirits, and their most solemn rites were held in groves of sacred oak trees. They gathered mistletoe from the oak, cut the oak down, and used the log to create a sacred fire to honor Yaroal, the Celtic god of fire. That was the origin of the Yule log many people still burn on Christmas. It is an offering to Yaroal to convince him to use his power over fire to end the winter quickly.

There are more recent and immediate examples of how the oak figures in human umwelts. In the umwelt of a logger the oak is nothing but wood, one of many trees, nothing special. To a conservationist the oak is unique, a treasure to be preserved. To a child lost in the woods at night the oak might be a terrifying magical object inhabited by gnomes and spirits. The knobby bark might seem a face in the moonlight, full of threat and mischief. And finally, to an aerial geographer flying over the oak and taking photographs to use in map making, the oak is hardly noticed at all. A minor feature of the whole landscape.

The oak exists in many umwelts. As von Uexkull says in his essay:

In all the hundred different umwelts of its inmates, the oak tree as an object plays a highly varied role, at one time with some of its parts, at another time with others. Sometimes the same parts are large, at others they are small. At times its wood is hard, at others soft. One time the tree serves for protection, then again for attack.

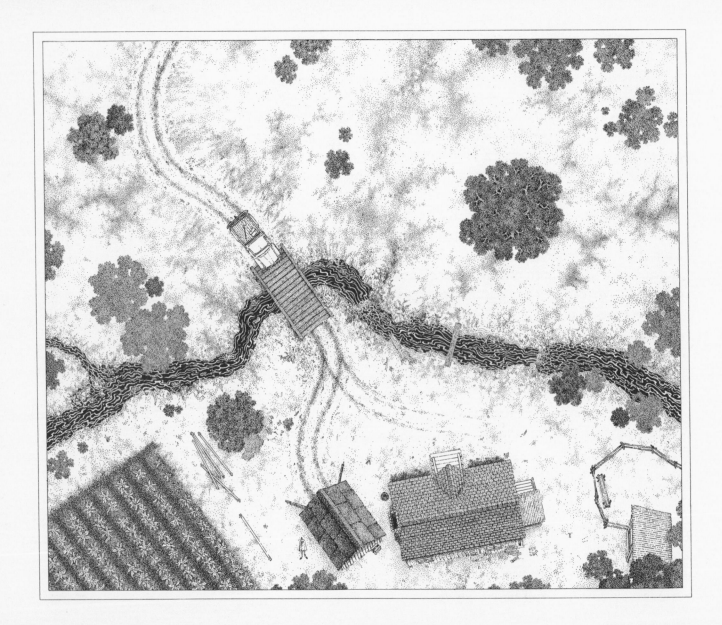

But what about the real oak tree? Is it big or small? Nothing special or a unique treasure? Menacing or protective? Is its wood hard or soft? To demand unambiguous answers to these questions is to make the world simpler and less interesting than it is. Why not give up worrying about the real object and rephrase the questions: In whose umwelt is the oak big and in whose is it small? Which animals find oak hard and which find it soft? For whom is the tree unique and for whom is the oak common? Who is menaced by the oak and who is protected by it?

By asking these questions we point ourselves toward observation and discovery. We actually know very little of the world, even of what surrounds us every day. It is worth taking the time to think of the variety of ways in which the environment could be structured and to discover how different animals actually structure it. We are all collections of atoms, specks in the universe, just the right size in our own worlds, giants to fleas, midgets to whales. The human view of the world is only one of many. It enriches our understanding of ourselves to move away from familiar worlds and attempt to understand the experience of other animals with whom we share common ancestors. The respect for other forms of life we can gain from these efforts might in some small way help us work toward preserving the world we share.

If You Liked This Book

Here are four books you might enjoy reading. They are all interesting and available in book stores and libraries.

The Magic of the Senses by Vitus Dröscher, Harper & Row, New York, 1971. If you are interested in learning more about the senses of animals and how they work, *The Magic of the Senses* not only has clear descriptions of sight and hearing, smell, taste, and pain but chapters on animal navigation, migration, gravity, and the electrical sense.

Animal Architecture by Karl von Frisch, A Helen and Kurt Wolff Book, Harcourt Brace Jovanovich, New York, 1974. This is a fascinating book about the structures built by insects, fish, reptiles, birds, and mammals and about the animals who build them. It is filled with hundreds of beautiful drawings and photographs.

King Solomon's Ring by Konrad Lorenz, Thomas Y. Crowell Company, New York, 1952. Lorenz has written several books you might enjoy. We've selected this one because it has interesting and amusing observations and anecdotes about animals most of us can observe easily—dogs, fish, hamsters, birds. It might make you want to watch what goes on in a fish tank for hours on end, and there is even a chapter on what Lorenz thinks are ideal pets.

The Curious Naturalist by Niko Tinbergen, A Doubleday Anchor Book, Garden City, New York, 1968. This book is not only about birds and insects, it is also about how several naturalists observed animals in their natural habitats and made discoveries about their behavior.

Index